WHY BIODIVERSITY MATTERS

All life on Earth has the right to exist, but as we teeter on the verge of a sixth extinction this book discusses why biodiversity matters and why we should care if species go extinct.

We are witnessing the largest and fastest rate of extinction in the history of the planet. While the concept of rights is a human one, all plants and animals strive to survive, and this book argues for their rights to continue doing so without being driven into premature extinction by human actions. Acknowledging and describing the practical reasons for conserving biodiversity, this book argues that these should not overshadow the compelling ethical reasons to care about the future of species other than our own. However, the issues are complex. What do we do when faced with an immediate ethical choice where biodiversity rights, animal rights, human rights, economic development and ecosystem survival all get mixed up together? There are seldom hard and fast answers, but thinking about and understanding a variety of points of view will help us make informed trade-offs. Drawing on his vast practical experience, the author presents insightful perspectives and real-world examples with the hope that this book will instigate a much-needed rethink about why and how we practise conservation.

This book is essential reading for all those concerned with sustaining our planet, and all who inhabit it, in the face of climate breakdown, biodiversity loss and ecological collapse.

Nigel Dudley is a consultant ecologist who has worked with international organisations, including WWF International, IUCN and UNESCO. He is Co-founder of Equilibrium Research and Industry Fellow in the School of Earth and Environmental Sciences at the University of Queensland, Australia. He is the author/editor of numerous titles, including Co-author of *Leaving Space for Nature* (Routledge, 2020), *Arguments for Protected Areas* (Routledge, 2010) and *Authenticity in Nature* (Routledge, 2011).

Changing Planet

Changing Planet publishes books addressing some of the most critical and controversial issues of our time relating to the environment and sustainable living. The series covers a broad spectrum of topics, from climate change, conservation, and food, to waste, energy, and policy, speaking to the varied and pressing issues that both human and non-human animals face on our planet today.

The Avocado Debate
Honor May Eldridge

Why Biodiversity Matters
Nigel Dudley

For more information about this series, please visit: www.routledge.com/ Changing-Planet/book-series/CPL

WHY BIODIVERSITY MATTERS

NIGEL DUDLEY

Routledge
Taylor & Francis Group

LONDON AND NEW YORK

Designed cover image: © Getty Images

First published 2024
by Routledge
4 Park Square, Milton Park, Abingdon, Oxon OX14 4RN

and by Routledge
605 Third Avenue, New York, NY 10158

Routledge is an imprint of the Taylor & Francis Group, an informa business

British Library Cataloguing-in-Publication Data
A catalogue record for this book is available from the British Library

ISBN: 978-0-367-35591-3 (hbk)
ISBN: 978-0-367-36520-2 (pbk)
ISBN: 978-0-429-34667-5 (ebk)

DOI: 10.4324/9780429346675

Typeset in Joanna
by codeMantra
Printed and bound by CPI Group (UK) Ltd, Croydon, CR0 4YY

CONTENTS

ACKNOWLEDGEMENTS

This work has drawn from countless conversations with friends and colleagues over many years. I can't thank everyone, so I'm simply listing those who have been involved in the main case studies given here. Thanks to Wale Adeleke, Mike Balzer, Mike Belecky, Sarah Brooke, Harri Karjalainen, Stewart Maginnis, Stephanie Mansourian, Vinod Mathur, Sean Maxwell, Brent Mitchell, John Morrison, Jeff Parrish, Khalid Pasha, Gert Polet, Madhu Rao, Hannah Scrace, Nik Sekhran, Luis Neves Silva, James Watson and Liza Zogib. They may not, probably will not, agree with everything herein.

Three friends and colleagues passed away during the period in which this book was written. Norman Myers died in 2019 after a long illness. His pioneering writings on tropical forests made me drop what I was doing and focus on issues of forest loss, which have remained a major part of our work ever since. Eleanor Sterling died in early 2023, a brilliant ecologist who cared deeply for the communities she worked with, in many countries, and did a huge amount to bring together the various strands of human and biodiversity rights with which this book is concerned. And Kathy MacKinnon died unexpectedly in March 2023, still working flat out for the conservation movement to which she dedicated her life and leaving a huge professional and personal hole in many of our lives. I am grateful to them all and miss their presence and wise counsel.

For many years, I've worked with my partner Sue Stolton on Equilibrium Research, more recently we've been joined by Hannah Timmins. Special thanks to both of them, including for their comments on the manuscript. My 30-year personal and professional partnership with Sue explains why the text varies between "I" and "we" when discussing personal examples. I am also very grateful to Kent Redford and Stephen Woodley for reading and commenting on the draft. And finally, thanks to Katie Stokes at Earthscan for patience beyond the call of duty and for her own inputs to the following manuscript. As ever, mistakes and opinions are my own responsibility.

1

INTRODUCTION

DO SPECIES MATTER? LAYING OUT THE CASE

A mass of excited people cluster around the edge of Hoàn Kiếm Lake in Hanoi old town, Vietnam. The evening light is punctured by camera flashes, people are laughing and pointing; it's a bit like an impromptu party. A puzzled German tourist asked me what all the fuss is about. I tell her that someone has spotted Cụ Rùa, *Great Grandfather Turtle*, a massive soft-backed turtle which has lived in the lake for as long as anyone can remember, swept in with flood water from the Red River many decades ago or maybe even deliberately introduced. For many Taoists, Cụ Rùa was considered sacred, linked to stories of a giant turtle that stole a magic sword from Emperor Lê Lợi in the fifteenth century and returned it to the Golden Turtle God. The name Hoàn Kiếm means *The Lake of the Returned Sword*. Long considered to be a myth by biologists, the giant turtle was rediscovered in the lake in the 1960s, when there were at least two remaining. One was killed by a fisherman and its body is preserved at a lakeshore temple, visited by hundreds of people every day.

To biologists Cụ Rùa had a different significance because it was "functionally extinct", being one of the last – perhaps even the last – individual of a species of the genus *Rafetus* that has no realistic chance of survival. The English language doesn't really have a word for "last in the line" in this sense although "ending" has been suggested, used as a noun.[1] There is currently a debate about whether the Hanoi turtle, *R. leloii* (sometimes also referred to patriotically as *R. vietnamensis*), is different enough from the Yangtze giant softshell turtle *R. Swinhoei*, to warrant being recognised as a separate species. Here science may be getting mixed up with Chinese-Vietnamese political rivalry. Either way it is or they are almost certainly doomed to extinction

DOI: 10.4324/9780429346675-1

because even if the two species are really one, there are at most five known individuals still alive, mostly in zoos, and all considered too old to reproduce. Once common on the sand banks of the great rivers of China and Vietnam, the turtles have been overhunted and, literally, had the ground cut from under their feet by dredging operations and the canalisation of major waterways. Scientists are still hunting for other individuals, but with no luck so far.[2] Cụ Rùa, a female, ended her long life placidly feeding alone in a muddy, polluted lake in a capital city. Like many others, I've taken some fuzzy photographs, surrounded by the noise of roaring motorbikes, groups of people doing Tai Chi on the banks of the lake, couples posing for wedding photographs and tourists drifting by. It's a strange place to be doing field work.

Our generation and the generations that immediately follow are witnessing the largest and fastest rate of extinction in the history of the planet. There is nothing inherently wrong with extinction: the fossil record tells us that all species disappear, outcompeted by stronger, more successful rivals, isolated and eventually killed off by changing weather conditions or evolving into something else. There is no reason to suppose that Homo sapiens will be different, and science fiction writers have had fun imagining what we might evolve into.[3] But the rate and scale of extinction is unprecedented[4] and risks disrupting whole ecosystems. Our immediate descendants are likely to be living in a world abruptly far less diverse, and far less stable.

Any active ecologist will continually see evidence of irreparable loss. In 2001, Sue and I stood with Vinod Mathur in Keoladeo National Park in Rajasthan, India, watching the last two Siberian cranes (Leucogeranus leucogeranus) that had managed to make the dangerous migration south across Central Asia to what had been their winter-feeding grounds for at least 500 years. On a misty evening, they were just a pair of small dots on a wet meadow. The sight of two isolated birds where there should have been a flock of hundreds was extraordinarily poignant; cranes are social birds with complex interactions, and we can assume that lonely birds really do feel lonely. Next year there were none,[5] the last individual probably making a feast for a hunter and his family somewhere over Afghanistan, and none have been seen there since. The whole species remains critically endangered.[6]

This short book is about biodiversity and why it matters. It is written in an attempt to try to reconcile some very different world views, including within the conservation movement, and to counter a growing tendency to pigeon-hole nature conservation into something that is anti-people, elitist and an unaffordable luxury in a planet under such environmental and social pressure. Whilst acknowledging and describing the practical reasons for conserving biodiversity, I argue that these should not overshadow, far less replace, the compelling ethical reasons to care about the future of species other than our own. To do so I am going to suggest a definition of "biodiversity rights" as an ethical concept distinct from, albeit constantly overlapping with, other ethical positions, particularly human rights and animal rights. These different concepts often rub up against each other and can potentially clash; digging down into some of these practical issues will take the largest section of the book. Then I summarise how some other stakeholders see these issues and look at the ways in which they have been addressed by individuals, governments and the international community. I'm coming at this from the perspective of a comfortable, white male, steeped in western thought and tradition; I am aware of the limitations that this imposes. There is a rich literature already to draw on; any ideas presented here have benefitted from much hard thinking by other people as well as my own experience and that of my friends and colleagues. Finally, and with some trepidation, I'll make some practical suggestions for how biodiversity rights can be put into practice without undermining other valid ethical considerations. But before starting the main text, I will give examples that illustrate two extremes of a complex debate.

LICHENS VERSUS JOBS

A crowded village hall in a small logging town somewhere in eastern Finland, close to the Russian border in the late 1990s; I'm listening to a discussion between representatives from the local timber industry and a group of conservation activists from Helsinki. The local people want to log an area of old-growth forest, and the conservationists want the forest to be protected: a scene played and replayed in hundreds of communities wherever people make jobs and money out of natural forests. The loggers have been strategic. Their spokesperson is a young woman from the community; she's smart, articulate and reasonable, pointing out the long tradition of forestry in the area and the impacts that any further protection will have on jobs and arguing that they will in any case be using sustainable practices. In deference to us visitors, the discussions take place in English. When it is the turn of the activists, their male speaker seems a little flustered and is much more emotional, his opening gambit is to ask: "What are we talking about jobs for? There are LICHENS at risk!" Whether speaking English or Finnish, the two sides were definitely not talking the same language.

I never found out how that particular argument played out, and in the years since mechanisation has got rid of many more jobs in the Scandinavian forest industry than any conservation group. And I have used the example of the hapless activist to illustrate quite a few things over the years, from the need for stakeholder engagement, trade-offs and a landscape approach to an example of how not to approach conflict resolution. But did he have a point? Take the argument to an extreme: if logging that particular forest were to cause the global extinction of a unique lichen species, would the temporary jobs created be worth the cost? Who decides? And on what grounds should these decisions be made – practical, political, ethical or cultural? Do human needs always predominate (and in which case *whose* needs)? Are the people immediately living in an area always the ones to decide? If so, where does that leave the rest of the human community?

DEATH OF A RHINO

These questions are not confined to obscure organisms like lichens. In 1989, the naturalist George Schaller was working in Vietnam when news broke that a poacher had been caught trying to sell the horn of a Javan rhino (Rhinoc-eros sondaicus), which until then had been assumed extinct on mainland Asia since the 1960s, the last individuals killed for food by the Viet Cong[7] as they fought a desperate war against the Americans. Schaller and his colleagues investigated and to their amazement found evidence of a small but still viable population; they estimated (or as George Schaller said to me "guesstimated") that there might be 10–15 individuals remaining,[8] a perilously small num-ber, but stable rhino populations have been rebuilt from less in Africa.

Steps were taken to protect the rhino, particularly through the creation of Cat Tien National Park, which covered the area believed to be its strong-hold.[9] In 1995, villagers reported witnessing a birth of a young rhino, but there was mounting poaching pressure and evidence of further losses. By 2000, when Mike Balzer was commissioned by WWF to do camera trap-ping for rhino in Cat Tien, he became convinced that there was only one individual left. In 2010, Sarah Brook and Simon Mahood lived in the pro-tected area for months and used tracker dogs to locate rhino droppings that were subject to DNA analysis to try to build a definitive picture of popula-tion levels, and the evidence from this was also that only a single animal remained. Halfway through their survey a corpse was found; it had been shot and the horn removed. For all intents and purposes, the Javan rhino is now extinct in mainland south east Asia. Furthermore, the rhino was genetically so different from the subspecies which continues to hang on, again at a critically low-population level, on the island of Java, that it may have virtually constituted a different species in its own right, making it by far the most significant extinction in the twenty-first century.[10]

What went wrong? In late 2010, WWF commissioned us to do an analysis of the combination of pressures that had driven the Javan rhino to extinction in Vietnam,[11] in contrast to situations where similar tiny groups of related rhino species had rebuilt viable populations in India and South Africa. We found that a combination of conditions had created a unique set of pressures that proved too much for the few remaining individuals to survive: high levels of poaching and the increasing sophistication of wildlife crime in Vietnam; corruption and weak enforcement despite years of capacity building by WWF; and diminishing habitat due both to military defoliants during the American War and increasing human migration.[12] Defoliants followed by logging[13] also created conditions for impenetrable scrub to develop that made things easy for poachers and hard for park guards (unlike the open savannah of Africa). Furthermore, political control was split between the Ministry of Agriculture and Rural Development and three local provinces, with the national park being a weak partner compared with the other actors. There was also, significantly in the present context, prevarication for a decade about plans to resettle two communities to provide space for the rhinos, and confusion between development and conservation projects, with some working in opposite directions.[14] There was little discernible support for the rhino amongst local communities. Development funds to help settlers, including Christians from northern Vietnam who had migrated south after the war ended, helped establish cardamom farms but fragmented viable rhino habitat.[15] In my earlier visits to the park, when we were trying to establish a follow-up to WWF's long-running project there, the local government was clearly split between a desire to save the rhino, fears that the species was doomed anyway and efforts to save it would create pointless human suffering, and the pressing needs of many human constituents. In the end, prevarication went on too long, causing one of the most significant wildlife losses of the last fifty years. As a conservation biologist I am frustrated that the government did not do more, but I'm fully aware of the complex choices facing decision-makers.

CHANGING PERSPECTIVES IN CONSERVATION

There is no consensus, even within the ranks of professional conservationists, about where to draw the line between the needs of plant and animal species and ecosystems and rights of human individuals and societies. Perhaps this is inevitable, but more worrying is that we haven't even agreed a common starting point. While planning out this book, an email came through on one of the list-servers run by the International Union for Conservation of Nature (IUCN). A friend circulated a paper arguing against what the authors believe is an overemphasis on conserving nature for the benefit of humans at the expense of calls to protect nature for its own sake. I was depressed but not surprised to see kickback from a number of people associated with IUCN or its commissions, arguing in some cases that they couldn't see that biodiversity – nature, species, plants and animals: whatever term you want to use – had *any* value at all except for what it provided to humans. I say not surprised because it isn't the first time I've heard IUCN staff and commission members present a strictly utilitarian worldview of nature (I'm not suggesting that this is a view shared by all or most people associated with IUCN). But I'm depressed that even people involved in a nature-conservation organisation should recognise no intrinsic value in nature. We need to have the discussion outlined in the following pages. Or rather, we need to look at existing discussions that have been taking place amongst philosophers of biodiversity and rights groups and inject them into a much broader policy debate.

This utilitarian view is gaining ground. Conservation International (CI) was set up in the 1980s by renegade scientists from The Nature Conservancy in a now infamous meeting at the Tabard Inn in Washington. Sponsored by the Moore Foundation, for a few years CI was probably the most biodiversity-focused of the "BINGOs" – Big International Non-Governmental Organisations – growing rapidly to run dozens of projects focused on their self-identified "biodiversity hotspots". Then, around a decade ago, there was an abrupt shift of mission; instead of focusing on biodiversity for its intrinsic values there was a very conscious decision to focus on biodiversity in the context of its usefulness for humans. There were ructions; many senior staff left. The new priority is captured by a series of quotes from the CI website:

> Since 1987, we have been fighting to **protect nature for people**... Conservation International works to spotlight and secure **the critical benefits that nature provides to humanity**... Conservation International empowers societies to responsibly and sustainably **care for nature, our global biodiversity, for the well-being of humanity**.

I know CI quite well, have worked for them in Madagascar, visited the US offices many times and admire their work. I don't actually think the mission shift has made much difference on a day-to-day basis; if you adopt the perspective that all biodiversity is important for humanity then the mission to conserve it is much then same as if you are driven by an intrinsic concern for other life. I'm guessing that most CI staff see both intrinsic and utilitarian values in nature. And yet … viewing biodiversity mainly or solely through the lens of human needs is the same as the old (and now largely abandoned) Christian approach of viewing all nature as being provided solely for our use by a God who made us in his own image. This view isn't confined to CI and I'm picking them out here as an example because they made the switch so publicly and explicitly. But it will eventually have ramifications on the way that we practice conservation, for example shifting concern away from obscure species that have no apparent "value" to humans, and deserves a more in-depth discussion than has been the case up until now.

This isn't the only side of the story. All around the world, communities, municipalities, research groups, religious groups, human rights groups, political parties, the European Union, the G77 plus China, town councils and groups of schoolchildren have all proposed formal processes for recognising the rights of nature. Most significantly, in December 2020, in the middle of the Covid-19 pandemic and after I started writing this book, the UN General Assembly adopted Resolution A/RES/75/220 on *Harmony with Nature*, which makes some of the strongest statements yet about the rights of nature. I'll discuss these important signs of progress later. But working on the frontline of conservation, I still see things going too far in the other direction. It is not that the benefits to humans are unimportant or that making the case is unnecessary; far from it and I'll outline these below. Utilitarian values are critical and they can also help to convince people who have no philosophical problem with extinction that conservation is important, but these shouldn't be the only arguments. Utilitarian views and ethical views push us in the same direction, but promoting one without the other can be dangerous. We have all the pieces on the board already, we just need more of an effort to see them properly integrated into the way we do business.

This book hasn't proved easy to write. I thought I'd be fleshing out an earlier set of arguments made before[16] and putting the issue into a wider context. But the more I thought and the deeper I dug there were so many subsidiary questions, complications, digressions that the whole thing has taken way longer than I expected and I thank the publishers for their patience. It's also opened up quite a few questions that I can't answer.

One of the reviewers of the original proposal for this book pointed out in no uncertain terms that I'm not a professional philosopher, with the inference that I really had no right to go meddling in things I didn't know much about. They make a very good point and I'm nervous about stepping so far out of my usual area of work. But I am also minded of the wise words of Leszek Kołakowski, who said *a modern philosopher who has never once suspected himself of being a charlatan must be such a shallow mind that his work is probably not worth reading* on the grounds that *of the questions that have sustained European philosophy for two and half millennia, not a single one has been answered to general satisfaction.*[17] I'm not going to attempt to solve them here either. But I have found too much of the literature about the philosophy of biodiversity and conservation to be inward looking, tied up in irrelevant semantic arguments and complacent in light of the urgent problems we face. Yet if we put aside the big questions raised by Socrates and his associates, there have been some genuine steps forward in terms of moral philosophy, which is principally what concerns us here and we need to understand and build on these in very practical ways.

To allow some space for the discussion that follows I am going to make some assumptions: that concepts such as "ethics" and "rights" are valid for instance and that non-human animals and also plants have certain rights, before applying these more generally to biodiversity (itself a concept that has spawned dozens of books).

In the mass of reading and talking I've done for this book, one of the problems with professional environmental philosophy is that it tends to stay within the profession. Very few of the practical or academic conservation biologists I know engage in debates about the philosophy of conservation on anything but a fairly superficial level. And as far as I can see few environmental philosophers attempt the messy business of trying to get conservation done. There are, of course, some honourable exceptions to both these statements. Many of us can only give a fairly vague explanation of why we engage in the practice of conservation at all, and I suspect most are driven mainly by emotional or spiritual ideas rather than carefully thought-out logical positions. Maybe this doesn't matter. But conservation is increasingly challenged, by human rights groups, animal rights advocates and by people who believe that human development and economic growth almost invariably outweigh the rights of an ecosystem to exist. If we are to defend our position, we'd better know why we hold it in the first place.

Philosophical musings are all very well, but what do we do when faced with an immediate ethical choice where human rights, economic development, ecosystem survival and a multitude of different individual needs and wants all get mixed up together? That is the kind of situation that anyone working on practical conservation is likely to be facing on a daily basis. What might seem to be an obvious truth to a conservation biologist may need to be backed up by more than emotion and self-righteousness. There are seldom hard and fast answers but thinking about and understanding a variety of points of view can help us make informed trade-offs.

As noted, there is already a considerable literature on the ethics of conservation by people who have spent their professional lives wrestling with these questions. What I'm principally interested in here is not the minutiae of the philosophical arguments but how we navigate them practically either as conservation professionals or as people working in other fields who are faced with complex issues relating to environmental management. How do we square the fact that much of the best-conserved land in Africa is set aside for trophy hunting, for people who are prepared to spend a very large sum of money to kill a spectacular animal (and be photographed sitting triumphantly beside the corpse)? What is someone who has devoted their life to protecting elephants to do when numbers increase to the extent that people in surrounding communities are being killed? How do we balance conservation of one species with the widespread killing of another – for example an invasive species? How do we react when the tiger we've struggled to conserve kills a villager? When if ever is it justifiable to force a human community to move to favour a non-human species? These are the kinds of questions that we are asking on a day-to-day basis and we don't always come up with a very coherent response.

There is often not a black and white, right and wrong. Instead, we usually face a complicated set of circumstances that needs to be addressed on a case-by-case basis. The book is by way of an introduction to some tricky issues. In the following pages, I include a lot of examples from personal experience, because real life is always trickier and more random than theory and thought experiments. I hope that this extended essay provides some practical advice. But when it comes down to it we are often on our own.

NOTES

1 Webster, R.M. and Erickson, B. 1996. The last word? *Nature* **380**: 386.

2 Van Pham, T., le Duc, O., Bordes, C., Leprince, B., Ducotterd, C., et al. 2022. Female wanted for the world's rarest turtle: Prioritising areas where *Rafetus swinhoei* may persist in the wild. *Oryx* **56** (3): 396–401.

3 For example, *Last and First Men* (1930) by the philosopher Olaf Stapleton, *Slan* by A.E. van Vogt (1940), *More than Human* by Theodore Sturgeon (1953), *Wolfbane* by Frederick Pohl and C.M. Kornbluth (1959) *Hothouse* by Brian Aldiss (1962); the *Lilith's Brood* trilogy by Octavia Butler (2000) and Marvel Comic's iconic X-Men series; there are literally thousands of examples. While many are simplistic or even banal, they helped to bring home to a generation of pulp science fiction readers that humans were not necessarily the ultimate pinnacle of the Earth's family tree.

4 Wagler, R. 2012. The sixth great mass extinction. *Science Scope* **35** (7): 48–55.

5 Vardhan H. 2002. Siberian Crane autumn migration 2002. Central Flyway. India. Crane Working Group of Eurasia Newsletter 4–5:29–30.

6 Mirande, C.M. and Ilyashenko, E.I. 2019. Species review: Siberian Crane (*Leucogeranus leucogeranus*). IUCN SSC Crane Specialist Group – Crane Conservation Strategy. In: *Crane Conservation Strategy*. International Crane Foundation, Baraboo, Wisconsin.

7 Dai, L.C. 2004. *The Central Highlands: A North Vietnamese Journal of Life on the Hồ Chí Minh Trail 1965-1973*. Thế Giới Publishers, Hanoi.

8 Schaller, G.B., Dang, N.X., Thuy, L.D. and Son, V.T. 1990. Javan rhinoceros in Vietnam. *Oryx* **24** (2): 77–80.

9 Polet, G., Van Mui, T., Dang, N.X., Manh, B.H., and Baltzer, M. 1999. The Javan Rhinoceros, *Rhinoceros sondaicus annamiticus*, of Cat Tien National Park, Vietnam: Current status and management implications. *Pachyderm* **27**: 34–48.

10 Brook, S.M., Dudley, N., Mahood, S.P., Polet, G., Williams, A.C., et al. 2014. Lessons learned from the loss of a flagship: The extinction of the Javan rhinoceros *Rhinoceros sondaicus annamiticus* from Vietnam. *Biological Conservation* **174**: 21–29.

11 Dudley, N. and Stolton, S. 2011. *Death of a Rhino: Lessons Learned from the Disappearance of the Last Javan Rhinoceros in Vietnam*. A report for the WWF AREAS project. Equilibrium Research, Bristol.

12 Quy, V. 2005. The attack of Agent Orange on the environment in Vietnam and its consequences. In: *Agent Orange and Dioxin in Vietnam 35 Years Later: Proceedings of the Paris Conference Senate*, 11–12 March 2005. France-Vietnam Friendship Association.

13 Thuat, N.H. and Mai, Y.H. 2013. Cat Tien National Park. In: Sutherland, T.C.H., Sayer, J. and Hoang, M.H. (eds.) *Evidence-Based Conservation: Lessons from the Lower Mekong*. Earthscan from Routledge: London and New York 17–28.

14 MARD. 2003. *Reasonable Resettlement Project for Cat Tien National Park*. Centre for Consultancy Investment and Construction, Ministry of Agriculture and Rural Development, Hanoi.

15 Brook, S.M., et al. 2014. Op cit.

16 Dudley, N. 2011. *Authenticity in Nature: Making Choices about the Naturalness of Ecosystems*. Earthscan, London.

17 Kołakowski, L. 1988. *Metaphysical Horror*. University of Chicago Press, Chicago, IL.

2

THE CONCEPT OF BIODIVERSITY

The word "biodiversity" didn't even exist when I started to work as a conservationist. It was first coined in 1986, as a conflation of "biological diversity" by the ecologist E.O. Wilson during a conference organised under the auspices of the National Academy of Sciences in the United States.[1] Seven years later, at the Earth Summit in Rio de Janeiro, biological diversity itself gained important recognition when the Convention on Biological Diversity (CBD) came into being. Ironically, given the central role of American scientists in defining biodiversity, the US government is joined by the Vatican City as the only nations that have never signed the CBD.

Biodiversity is a summary term that describes the diversity of life: ecosystems, species and genetic variation within species. This means that biodiversity encompasses *objects*, like an orchid or a microbe or a particular subspecies of tiger, but also *interactions* including the myriad ways in which species connect to each other within ecosystems. Many aspects of ecosystems are temporal, changing over time, so that biodiversity is properly represented in four dimensions. Species, genetic diversity and ecosystems are important, although most people, if they understand the term at all, tend to focus on species. As with virtually everything discussed here this definition isn't accepted by everyone and there is a vigorous debate in the philosophical literature about whether the word "biodiversity" means anything, whether it is useful and if it should be abandoned ("eliminativism").[2] Whilst acknowledging this discussion, ecologists generally know what they mean when they use the term and we have a whole UN convention based around its management so I will continue to use it here.

DOI: 10.4324/9780429346675-2

Broadly speaking a *species* has most commonly been distinguished from another species if the two cannot interbreed and create a viable offspring. But it gets a bit more complicated. We've already seen from the rhino example discussed above that the boundaries are sometimes fuzzy and experts are still arguing about whether the mainland rhino, until recently found in Vietnam, was genetically distinct enough from the animals found on the island of Java to constitute a separate species. They were at the very least recognised as Evolutionary Significant Units, which should be treated as distinct.[3] Even finding out if two widely separated populations of what might or might not be the same species can interbreed has lots of practical challenges. In fact, there are over 70 definitions of a species of which 48 are generally accepted by natural scientists,[4] based on criteria such as ecological niche, physical characteristics and – most importantly – DNA sequence. DNA evidence provides the most concrete indicator. It has been suggested that a difference of as little as 2 per cent in DNA can constitute a different species, leading to a vast increase in the number of species. The field guide to African birds I bought recently has a lot more species in than the one I had when I was just starting out. We now recognise two African elephant species whereas we only had one before[5] and lemur numbers doubled to around a hundred species over a period of 20 years. This has also led to the recognition of "cryptic species"; organisms that look the same but are genetically distinct; DNA barcoding showed that the neotropical skipper butterfly *Astraptes fulgerator* in Costa Rica was actually a complex of ten species.[6] Species complexes have been recognised for much longer among plants, with the blackberry or bramble (*Rubus* spp.) being a well-known example. Then in animals there is the phenomenon of "ring species" – a series of subspecies that have evolved over time, all of which can interbreed with each other, but where the subspecies at either end of the line of dispersal cannot interbreed. Within fungi, themselves only recognised as being separate from plants a few decades ago, conditions become so abstruse that some mycologists are claiming the whole concept of a "species" to be unworkable.[7]

Not everyone is enamoured by the use of small genetic variation as the determinant of speciation. Taxonomists are colloquially divided into "lumpers" and "splitters"; those who have the tendency to assume that a single species can have quite wide variation and those who would instead prefer to identify a range of closely related species. We have three species of the crossbill bird in the UK, whereas when I was a boy we only had one, a distinction that many ornithologists roll their eyes about. This issue won't be sorted out any time soon. The entire concept of a "species" is a human construct, and an attempt to draw firm lines where they often don't exist in nature. DNA sequencing was supposed to settle arguments once and for all but this has not proved to be the case.

Although it can seem rather trivial to anyone outside the sometimes competitive world of taxonomy, these debates have important implications in terms of conservation policy, for instance in determining whether a species is eligible for protection under national conservation laws and policies. But from our perspective here the distinctions are less important, because biodiversity includes genetic variation *within and between* species, suggesting that the variation remains significant whether a group of organisms are recognised as one variable species or a bunch of closely related species. This genetic variation is important for providing evolutionary potential, increasing the chances of a particular species surviving sudden changes in environmental conditions or challenges – such as a newly emergent disease – and to maintain a healthy breeding stock. Small, isolated populations result in inbreeding and a gradual weakening of individuals, a phenomenon recognised for millennia and of critical importance in the management of protected areas and isolated, endangered species. Multiple factors can combine, leading to a spiralling decline sometimes known as an extinction vortex.[8] Once beyond a certain point, it becomes almost impossible to halt or reverse.

Just as important, but potentially much more difficult to quantify, is that biodiversity also includes *ecosystems*, the wider environment in which species co-exist. Biodiversity therefore does not just encompass objects, like species, but also the physical, chemical, temporal and behavioural interactions between them. An oak tree is not simply a being existing in a vacuum, it will be supporting hundreds of species that eat its leaves, and other species that feed on *those* species, it will in turn be supported by fungi that take some nutrients from the tree and replace others, and by squirrels and jays that bury its acorns and thus help to ensure new trees grow and spread geographically. Once dead the oak will support other species of fungi, insects that feed on dead wood, woodpeckers that nest in its trunk, and so on, a whole world in fact. Plants release toxins to dissuade herbivores, countless fungi and microbes influence every stage of the life cycle of an ecosystem, everything interacts and species exist in a complex and constantly changing pattern.

This means that, from a conservation perspective, it is not enough to isolate a species in a zoo, botanical garden or seedbank, valuable those such actions may be in emergency. Instead, to reflect the full range of biodiversity any one species needs to be interacting with many others, and also with soil and water and the rest of the physical environment. And to make matters more complicated, such interactions are constantly evolving and changing over time, in response to changes in climate and other environmental conditions. The old idea of ecosystems evolving to a "climax vegetation" pattern and then staying the same indefinitely has been replaced with a recognition that ecosystems are constantly changing as species come and go, climatic and environmental conditions alter and new food webs and other interactions develop. We still only understand a fraction of these interconnections.

Over time, concern about "saving biodiversity" has shifted from a focus on a few rare species to concern about whole ecosystems, including both those species we know about and those not yet described by science. This might seem obvious to anyone reading in the twenty-first century. But when groups like Friends of the Earth and WWF started campaigning against the loss of tropical forests in the 1980s, there were intense debates inside those organisations about whether or not the public would understand or care about something as abstract as an ecosystem. The campaigns were amongst the most successful run by either organisation, a clear example of the experts underestimating civil society.

This also suggests something about the limitations of the term "biodiversity" as a way of bringing wider society into the conservation debate. Although the word looms large in natural science and in the minds of Ministries of Environment the world over, it has yet to gain resonance with the public. When people were asked what biodiversity is in 2010 the commonest answer was that it is a kind of washing powder,[9] rather than a fundamental building block of conservation. Hopefully understanding has progressed a little since but it still isn't a term that creates a clear mental picture for most people. Phrases like "clean air" and "pure water" mean something to everybody, as does "wildlife", whereas if non-specialists recognise the term biodiversity at all, they are likely to interpret in terms of diversity of organisms, rather than also including the ecological interactions between them.

There is also uncertainty about whether concern about biodiversity extends to those living things that have been substantially modified by human endeavour: pets, livestock, crops, fruits and many garden plants. The debate becomes even more intense if synthetic biology is mentioned, with strong opinions about whether or not it is morally acceptable or safe enough to experiment with.[10] Could for instance scientists genetically modify mosquitoes to reduce the spread of avian malaria that is threatening endemic bird species with extinction?[11] And if so, would those "created" species have rights?

Purists would say no, but the IUCN Species Survival Commission has a specialist group dedicated to crop wild relatives *and landraces*,[12] my emphasis, looking at the survival of traditional crop varieties. Furthermore, what about cultural ecosystems, created by humans and sometimes of ancient lineage? In the European Mediterranean, species have evolved to managed landscapes over millennia, and without them some species would now have no-where else to go. People involved in conservation in Spain, southern France, Italy and the coastal Balkans conserve the ancient terraced landscapes for their conservation values.[13] Other anthropogenic ecosystems, like rubbish dumps that provide habitat for rats and cockroaches, attract less interest. While these distinctions make sense on a gut level, is there any basis for them apart from cultural norms? Post-industrial sites can be spectacularly rich in endangered species because they contain rare areas of disturbed and unmanaged habitat with a particular set of (usually temporary) conditions.[14] But they are generally ignored by conservationists.

The concept of biodiversity is therefore quite new and at first sight straightforward, but on a deeper look is revealed to be complicated, changeable and with disagreements about where boundaries lie. For our purposes, "conserving" biodiversity" can be interpreted as conserving species, at population sizes large enough to contain a wide range of genetic variation, in functioning ecosystems. Exactly what that means is also subject of debate, but one set of criteria for vertebrates suggests that successful conservation should include demographically and ecologically self-sustaining populations that are genetically robust, healthy, representative and replicated, with resilience across a range.[15] Achieving this in practice is no mean task.

NOTES

1 Wilson, E.O. (ed.) 1988. *Biodiversity*. National Academy of Sciences, Washington, DC.

2 Garson, J. Plutynski, A. and Sarkar, S. (eds.) 2017. *The Routledge Handbook of the Philosophy of Biodiversity*. Routledge, London. See particularly Chapters 3 and 6. Sarkar, S. Approaches to biodiversity, 43–55 and Santana, C. Biodiversity eliminativism, 86–95.

3 Fernando, P., Polet, G., Foead, N., Ng, L.S., Pastorini, J. and Melnick, D.J. 2006. Genetic diversity, phylogeny and conservation of the Javan rhinoceros (*Rhinoceros sondaicus*). *Conservation Genetics* 7: 439–448.

4 Nuwer, R. 2013. What is a species: Insight from dolphins and humans. *Smithsonian Magazine*. https://www.smithsonianmag.com/science-nature/what-is-a-species-insight-from-dolphins-and-humans-180947580/

5 Roca, A.L., Georgiadis; N., Pecon-Slattery, J. and O'Brien, S.J. 2001. Genetic evidence for two species of elephant in Africa. *Science* **293** (5534): 1473–1477.

6 Hebert, P.D.N., Penton, E.H., Burns, J.M., Janzen, D.H. and Hallwachs, W. 2004. Ten species in one: DNA barcoding reveals cryptic species in the neotropical skipper butterfly *Astraptes fulgerator*. *Proceedings of the National Academy of Sciences* **101** (41): 14812–14817.

7 Money, N.P. 2013. Against the naming of fungi. *Fungal Biology* **117**: 463–465.

8 Frankham, R., Ballou, J.D. and Briscoe, D.A. 2002. *Introduction to Conservation Genetics*. Cambridge University Press, Cambridge.

9 https://www.bbc.com/news/science-environment-11546289

10 Redford, K.H. and Adams, W.M. 2021. *Strange Natures: Conservation in an Era of Synthetic Biology*. Yale University Press, New Haven, CT.

11 Regalado, A. 11th May 2016. The plan to rescue Hawaii's birds with genetic engineering. *MIT Technology Review*.

12 https://www.iucn.org/our-union/commissions/group/iucn-ssc-crop-wild-relative-specialist-group

13 Atauri, J.A. and de Lucio, J.V. 2001. The role of landscape heterogeneity in species richness distribution of birds, amphibians, reptiles and lepidopterans in Mediterranean landscapes. *Landscape Ecology* **16**: 147–159.

14 Muratet, A., Muratet, M., Pellaton, M., Brun, M., Baude, M. et al. 2021. Wasteland, a refuge for biodiversity, for humanity. In: Di Pietro, F. and Roberts, A. (eds.) *Urban Wastelands*. Springer, New York: 87–112.

15 Redford, K.H., Amato, G., Baillie, J., Beldomenico, P., Bennett, E.L. et al. 2011. What does it mean to successfully conserve a (vertebrate) species? *BioScience* **61** (1): 39–48.

3

THE CONCEPT OF RIGHTS

A few years ago, I was shown round a "slave castle", one of around sixty slave trading centres built on the coast of in Ghana. There are a lot of mixed feelings about even making such places into tourist destination,[1] but that is not my focus here. At the time when Jane Austen was writing her novels, slavery was an accepted part of society; abolitionists were still regarded as a deluded and dangerous fringe. The consensus changes surprisingly fast. Attitudes to gender rights and gay rights have undergone a revolution in my lifetime. This doesn't mean that gender disparities have been somehow magically eliminated or that people no longer face challenges because of their sexuality (or indeed that slavery has disappeared from the world) but that there has been a sea change in attitudes which it seems unlikely, although never impossible, to be reversed.

THE EVOLUTION OF RIGHTS

The idea that people have rights evolved slowly through history. Attitudes in pre-literate society can, in large part, only be inferred. Early written frameworks come from two sources: moral teaching associated with the emergence of the world's great religions and the flowering of philosophy that took place in a small area of Greece a few hundred years before the emergence of Christianity. In South Asia, the emergence of Hinduism some 4,000 years ago was accompanied by moral teaching, such as right action as a component of dharma,[2] a powerfully argued moral framework, albeit without a strong philosophical explanation about why people should act in

DOI: 10.4324/9780429346675-3

a particular way. The centuries between 800 and 300 BCE saw an explosion of new religious and philosophical ideas and beliefs, including the teachings of Buddha, Confucius and Lao Tzu in the east but also of Socrates, Plato and Aristotle in the west, all providing moral and ethical frameworks and challenging many traditional values.[3] The three religions that emerged in the Middle East, Judaism, Christianity and Islam, also offered ethical advice to their followers, encapsulated by the Ten Commandments for the Jews, the Sermon on the Mount for Christians and the moral teachings of the Koran for Muslims. Most of these teachings were about what an individual should or should not do, either to live a moral life or to gain some future reward in the afterlife in the case of many religions; the distinctions become blurred in practice.

Progress has been slow; many scholars regard the whole history of western philosophy as a footnote of the thoughts of the Greeks two and a half millennia ago.[4] Aristotle's writing was lost in the west for centuries and preserved by Islamic scholars, so that the works eventually made their way back to Europe again and flowered in the Renaissance. Progress in ethical discourse is seldom smooth. Advances in moral thinking emerge gradually, ebb and flow, and continue to be resisted by some people long after the bulk of opinion has shifted. Nor is progress one-way, ethical attitudes can sometimes reverse, most spectacularly in the last century perhaps in the emergence of the systematic brutality embodied in the Holocaust, although sadly there are several other contenders. And even when an ethical position is accepted by society, plenty of individuals will flout the rules.

These early moral systems talked much more about *expectation* than about what an individual might expect in terms of *rights*, although the latter could sometimes be inferred. Universal rights were not generally considered and the poorest or weakest members of society enjoyed far fewer rights than those with wealth and power. All the Greek moral philosophers were comfortable with the concept of slavery for instance, and the secondary position of women in society, while Plato also opposed democracy and favoured rule by an elite. Rights as an explicit concept emerged much later, in the west for instance through early writing by St Paul, Christian thinkers such as St Augustine and the concept of natural law that emerged in the Medieval period and was taken up again during the Enlightenment.[5] The concept of universal human rights only became codified in the twentieth century, significantly at the signing of the Universal Declaration of Human Rights in 1949.[6]

The concept of rights emerges from the ethical positions taken by society. In the last few decades, the rights of women, children, gay people, transgender people and Indigenous peoples have all changed dramatically, both in law and more importantly in general attitudes within civil society. When I was at school in Britain older boys were allowed to cane (hit with a stick) younger boys who they considered had broken the "rules", gay people could be locked up in prison and there were virtually no women in positions of authority, indeed only a tiny proportion of women in any kind of professional position. These conditions (which persist in many countries) have undergone dramatic changes in large parts of the world. Other moral judgements remain much more disputed, like the shifting attitudes to abortion, pitching the rights of the mother against the rights of the unborn child, and the rights of people to move and live in different parts of the world.

The evolution of rights can be divided into a number of stages, many of which overlap. First, a few pioneers identify an issue and raise it, often through writing or oration: more recently also through film and social media. Such people are usually derided and not infrequently persecuted. Then the issue gradually gains credence with an increasing proportion of civil society and typically also by a number of politicians and other figures of influence, like the gradual acceptance of many of the ideas of feminism, even by people who still sneer at "feminists". Next there may be a statement or series of statements in support of the position, such as manifestos and sets of principles. In time these tend to come from increasingly mainstream sources. The statement from the International Labour Organization concerning *Indigenous and Tribal Peoples in Independent Countries*[7] is a good example of a previously minority issue moving into the mainstream. As a major step, rights are enshrined in law at national or international level, like the decriminalisation law on homosexuality in many countries. (Other countries are going in the reverse direction, like the recent introduction of the death sentence and life imprisonment for homosexual acts in Uganda, a sharp illustration of the ebb and flow of rights.)[8] If changes are to stick, there must be general acceptance by the mass of the population of a country or region. Progress is seldom uniform and for instance the Anglican Church is riven with bitter disagreements on issues like attitudes to gay people and the ordination of women priests. Rights come and go to a certain extent and it would be naïve in the extreme to assume that once gained, a particular set of rights can never be eroded again.

Until recently, all but a tiny proportion of the discussion about rights focused on human rights. Emerging mainly in the twentieth century, the focus gradually shifted towards other species.

RIGHTS FOR NON-HUMANS

Attitudes to the rest of the animal kingdom, and to plants, have generally differed between East and West, the Orient and the Occident, with religions in the former giving higher status to animals than in the monotheistic religions emerging from the Middle East. Whether they have made much difference in the day-to-day manner in which animals are treated is less certain and subject to debate.

Over much of the world, the mass of animals have been for most of history treated with a casual disregard, if not outright cruelty.[9] There are exceptions, including those species that have been kept as pets and others that become venerated by particular cultures or groups as sacred, where they may be treated with exaggerated respect. Dogs were first domesticated an estimated 35,000 years ago, long before crops and livestock,[10] in a process still imperfectly understood.[11] Many animals are regarded as sacred by one or more sectors of society,[12] including but not limited to many fierce predators.[13,14] Animals regarded as sacred may also continue to be hunted and killed in some situations. There are also sacred trees, both whole species imbued with particular value and individual, often old, specimen trees that have accrued certain cultural values, folktales and beliefs over the course of a long life.[15]

In the past, the distinctions made between animals and humans were often quite subtle, despite scripture suggesting otherwise. Many people in the West believed animals had souls for instance, although beliefs were much stronger in this regard amongst lay society than professional theologians.[16] In Renaissance thinking, birds were the highest animals, closer to humans than mammals or reptiles and in Christian iconography they were often intermediaries between God and humans.[17] Animals featured in folklore, religion and art, and in places where rulers could afford it the emergence of zoological gardens brought wealthier society into close proximity with a much wider variety of species than hitherto.[18] This nuanced approach was challenged in Europe by the philosopher Rene Descartes, who argued that animals were mere machines (the *beast machine*), incapable of thinking or feeling and operating wholly on instinct.[19] This more "scientific" attitude, although quickly challenged by other thinkers, ushered in a period of extraordinary cruelty. Early members of the Royal Society in Britain, one of the world's first scientific societies, carried out stomach-churning experiments in the belief that the howls of anguish from the luckless individual animals they used were not something to be taken seriously.

A more widespread and formalised belief that animals might have rights of their own came much later and was primarily a movement that developed properly in the early years of the twentieth century. But once it emerged, the animal rights movement flourished very quickly to become an important force for political and social change.

NOTES

1 Richards, S.L. 2010. Cultural travel to Ghana's slave castles: A commentary. *International Research in Geographical and Environmental Education* **11** (4): 372–375.

2 Knott, K. 1998. *Hinduism*. Oxford University Press, Oxford.

3 Armstrong, K. 2006. *The Great Transformation*. Atlantic Books, London.

4 Russell, B. 1946. *A History of Western Philosophy*. George, Allen and Unwin, London.

5 Vardy, P. and Grosch, P. 1999. *The Puzzle of Ethics*, 2nd edition. Harper Collins, London.

6 UN. 10th December 1948. Universal Declaration of Human Rights. Adopted by the UN General Assembly.

7 ILO. 1989. ILO Convention 169 concerning Indigenous and Tribal Peoples in Independent Countries. www.ilo.org/public/indigenous/standard/index

8 Okiror, S. 21st March 2023. Ugandan MPs pass bill imposing death penalty for homosexuality. *The Guardian*.

9 See for example Roberts, K. 1983. *Man and the Natural World: Changing Attitudes in England 1500–1800*. Allen Lane, Harmondsworth.

10 Galibert, F., Quignon, P., Hitte, C. and André, C. 2011. Toward understanding dog evolutionary and domestication history. *Compte Rendus Biologiques* **334** (3): 190–196.

11 Larson, G., Karlsson, E.K., Perri, A. and Lindblad-Toh, K. 2012. Rethinking dog domestication by integrating genetics, archeology and biogeography. *Proceedings of the National Academy of Sciences* **109** (23): 8878–8883.

12 For instance, Krishna, N. 2010. *Sacred Animals of India*. Penguin Books, New Delhi.

13 McKay, J.E., St. John, F.A.V., Harihar, A., Martyr, D., Leader-Williams, N., Milliyanawati, B. et al. 2018. Tolerating tigers: Gaining local and spiritual perspectives on human-tiger interactions in Sumatra through rural community interviews. *PLoS One* **13**: e0201447. DOI:10.1371/journal.pone.0201447

14 Stolton, S. and Dudley, N. 2019. *The New Lion Economy*. Unlocking the value of lions and their landscapes. Equilibrium Research, Bristol.

15 Barrow, E. 2019. *Our Future in Nature: Trees, Spirituality and Ecology*. Balboa Press, Bloomington, IN.

16　Thomas, K. 1983. *Man and the Natural World: Changing Attitudes in England* 1500–1800. Allen Lane, London.

17　Naïs, H. 1961. *Les animaux dans la poésie française de la Renaissance* : Science, symbolique, poésie. Marcel Didier, Paris.

18　Sahlins, P. 2017. *1668: The Year of the Animal in France*. Zone Books, New York.

19　Merchant, C. 1990. *The Death of Nature: Women, Ecology and the Scientific Revolution*. Harper, New York.

4

WHAT ARE BIODIVERSITY RIGHTS?

The main precept of this book is that all living species and ecosystems deserve thoughtful ethical consideration and treatment by humans. But defining exactly what those considerations are, how they relate to rights of other species including the rights of humans, and exactly what we mean by "deserve" is much more complicated. Rights of different groups and individuals are often at odds and we have little by way of guides to help us decide precedence or relative importance.

Biodiversity rights, as defined and used in this book, focus primarily on the rights of species, ecosystems and the evolutionary potential to continue to exist. The last clause implies that ecosystems are of sufficient size and well enough connected to other ecosystems to be sustained in the long term and that species contain enough genetic diversity to help adaptation to changing conditions over time. These is in contrast with a focus on the existence and other rights of individual animals ("animal rights") or of individual plants.

In more formal terms, this means a focus on ecocentrism or eco-holism – the moral rights of species and ecosystems, rather than biocentrism, generally interpreted to be the moral rights of individuals amongst species or sentientism, the moral right of sentient creatures.[1] The differences between these three will be examined in greater depth in the section on conflicts between different approaches to environmental ethics. There are a plethora of other ways of valuing biodiversity, most commonly instrumental value, broadly speaking the practical values that species have for us. This is critical but not our main focus here; an overview of the instrumental value of species and ecosystems will be given immediately following this chapter.

DOI: 10.4324/9780429346675-4

All these ideas have been around for a while and there is a rich supporting literature.[2] Some of the deep-rooted arguments about the validity of terms like "biodiversity" and "rights" have already been running for decades and won't be solved anytime soon. Interesting though they are, they are in danger of leaving us hanging without any answers.

It should be noted that some philosophers dispute the possibility of biodiversity having intrinsic value or moral standing because biodiversity doesn't have "interests" and is not a "thing", but rather an aspect of things.[3] Others disagree, saying that biodiversity can have value if it is intrinsically valued by ourselves,[4] which is a little circular and still links any values or rights of other species and of ecosystems back to our own attitudes. I'd go further and say that biodiversity is a collective term for many things and if we observe how these things (species and their varieties, and in a slightly more complicated way ecosystems) behave then we can infer their "interests", even if we are unable to ask them directly, as explained below. But this is a long-running debate and I am aware not everyone will agree.

Considering species and ecosystems rather than individuals makes definition of rights more complicated, because in the long term (usually stretching mainly beyond the timescale of human history), species either evolve or disappear and ecosystems themselves develop over time as environmental conditions change. If we take both *survival* and *evolution* as important components of the trajectory of living species, the core concepts of biodiversity rights as addressed in this book might be stated as:

THE RIGHT OF ALL SPECIES TO CONTINUE THEIR NATURAL SPAN OF EXISTENCE WITHIN A FUNCTIONING ECOSYSTEM

Each word is loaded with implications. I'll look particularly at *right*, *all*, *species*, *natural span*, *existence* and *functioning ecosystem*.

Right is a human concept. We have no way of knowing whether even the most intelligent animals understand the concept of rights in anything but very immediate terms, such as issues of distributional "fairness".[5] Assuming what individual members of a species want in more metaphysical terms, such as species survival and long-term evolutionary pathways, is clearly ridiculous, particularly as we only have very hazy ideas about that from a human perspective. There are also many shades of opinion, and I'm aware that not every biologist agrees with the following. But a mass of evidence from evolutionary, genetic and behavioural studies suggests that every species has evolved in ways to maximise the chances of *survival of the species*. Although we are unable to ask the opinion of a panda, or a beetle, or an oak tree, we might reasonably take as a working assumption that species are driven to continue their existence. This emphasis on the species rather than the individual is supported by the willingness of parents of many species to sacrifice themselves, or put themselves in danger, to defend their young. Whether it is a mother octopus starving to death to defend her eggs, or the meadow pipits in the hills above my house mobbing any cuckoo that threatens to parasitise their nest, or a lioness defending her cubs to the death, the pattern is similar. This is not a blind drive to sacrifice. A bird that loses its mate while nesting will sometimes abandon the fledglings, which makes evolutionary sense if there is very little chance of pulling off a successful brood. Species that lay large numbers of eggs usually exhibit no care beyond depositing eggs in a suitable place and rely on sheer numbers of young to carry forward the next generation. But the general principle that evolution favours the population rather than the species holds true.

Plants have also evolved numerous complex ways of ensuring the survival of the next generation. A tree blown down in a storm will divert its remaining energy into producing a bumper crop of seeds. In this case, the "right" can be reflected by us recognising, responding to and supporting this drive for the continuation of a species' existence. It is, inevitably, still a completely human construct but one that has taken due consideration of how our fellow species behave. The situation is slightly more complicated for ecosystems as discussed below.

All implies that we are talking about everything that is alive: fungi, plants and animals, charismatic animals that everyone knows and loves but also tiny, obscure microbes that no human has noticed exist. The better-known discussion of animal rights is more limited: it ignores plants and sometimes makes very different philosophical judgements in consequence. The large majority of people actively involved in animal rights also focus overwhelmingly on the rights of larger and more intelligent animals (*sentientism*), which has some logic, and in practice also on those species whose appearance or cultural links make them particularly appealing, which makes less sense. In biodiversity rights everything is important. The question of whether everything is equally important, or if some animals are "more equal than others" to quote George Orwell,[6] is more complicated.

Species is a complicated concept in itself, as discussed above, an attempt to form sharp distinctions around complex and constantly evolving groups of organisms. But it is a useful shorthand that most people more-or-less understand. It includes genetic variation, the third aspect of biodiversity, both major genetic variants within a particular species such as the case of the Javan rhino described above, and the infinite smaller variations that occur. If a species is to survive in the long term by implication, it will need a good variety of genetic material to help facilitate evolution. Small populations confined to protected areas or other "islands" are likely to suffer inbreeding and decline unless they can continue to have genetic interchange with other populations. So here the term embodies both a distinction between different groups of organisms and also the variety within a particular species.

Existence in this case refers to the existence of a species rather than existence of individuals within a species, although there are clearly links. Plants and animals live and die; from the long-term perspective it is the survival of the species that is important rather than the survival of the individual, and natural ecosystems continue to function because the large majority of plants and animals are killed and eaten before they can reproduce. Frogs can lay thousands of eggs every year and only a tiny fraction survive to adulthood; similarly, it has been calculated that an oak tree can produce 10 million acorns during its lifetime yet the population of oaks in a natural landscape remains more-or-less constant. In the case of very rare species, reaching the edge of extinction, the existence of each individual may become of paramount importance. Every one of the rhinos in Cat Tien National Park was important, and their death led to the loss of a unique sub-species. Most surviving wild tigers are now individually identified and recorded, a sure sign that a species is in big trouble.

Natural span implies that a species should be able to survive until it either evolves into a distinctly different species or becomes extinct as a result of natural factors: the two alternatives that we assume eventually face every species. We might state this as *the right of the global ecosystem to experience rates of species extinction no greater than the norm expected through natural evolutionary and climatic processes*, or to put it more bluntly, *the right of species not to be driven to extinction by our actions.*[7] This includes both direct actions, like hunting,[8] and much more indirect impacts like extinction risk from human-induced climate change, which may shorten the existence time of many species.[9,10]

Functioning ecosystem is also complicated. At an extreme it means that if the entire remaining global population of a species is confined to zoos or arboreta this doesn't meet our definition of biodiversity rights. However, captive populations capable of seeding new wild populations are a massive step up from total extinction, like the population of Père David's deer (*Elaphurus davidianus*) in the herd at Woburn Abbey and zoos like Bronx Zoo in New York, which saved the species from extinction and allowed reintroduction into the wild in China.[11] A simple way of defining a functioning ecosystem might be *wild species living in natural conditions*. But given the extent to which the world and its climate have already been changed by our own actions, talk about "natural ecosystems" is sometimes problematic. We can nonetheless identify the elements of a functioning ecosystem: composition of species, ecosystems and genetic variation; spatial variation of vegetation with respect to age, size, etc.; continuity of the ecosystem over time; disturbance patterns and life cycles; size, edges and connectivity; degree of fragmentation; and so on. Also included are the innumerable interactions between species through predation, parasitism, commensalism, through chemical stimulants and deterrents, pollination, seed spreading, etc. This implies that species have a viable population, with enough genetic variation to provide resilience in the face of changing conditions,[12] and will interact with other species and with the inorganic materials in their environment in ways that reflect long-term evolutionary patterns. In terms of animals, this also implies that social interactions within and between species remain reasonably intact.

This does not necessarily imply that the ecosystem has to be entirely "natural". I don't think that birds living happily in suburban back gardens are having their rights impeded, even if their habitat is full of exotic flowers and neatly cut lawns, although that depends to some extent on how the garden is managed. More urban nestlings die of starvation than their relatives out in more natural conditions because there are fewer of the high protein insects available.[13] Whenever we intervene to prevent a species from becoming extinct we change the ecosystem, whether it is by altering fire regimes, or putting up a roosting box for bats or nest boxes for swifts or relocating a highly threatened species to an island away from introduced predators. There are many circumstances where this is quite justifiable, but it does imply that a given ecosystem is not functioning perfectly at least for the species under consideration. Focusing attention on a single species or group also risks distorting ecosystem function to disadvantage other biodiversity,[14] perhaps in ways that are only poorly understood or unrecognised. The small, highly managed reserves typical of the more crowded and developed parts of the world often fall into this category. So, a fully functioning ecosystem is an ideal, but in its absence surrogate measures may be appropriate on a temporary basis. This in turn has major implications for the stewardship roles that humans play in ecosystems.

The argument that we can infer that species are acting in ways to ensure their intergenerational survival is trickier in the case of ecosystems, not least because the latter change all the time and every ecosystem is unique to a greater extent than every individual in a species. Maybe in these cases we simply need to make an ethical leap of faith and assume that ecosystems *should* have rights. As discussed in the chapter on responses, many faith groups and a number of governments have already done this without getting caught up in philosophical knots.

And finally, biodiversity rights apply to humans as well. We're a species amongst species and have rights ourselves. This becomes important when biodiversity rights (or perhaps in this case it is better to say rights of all non-human biodiversity) bump up against human rights, as discussed below. By including humans amongst other species as part of "biodiversity" the definition also comes quite close to the philosophy of some faith groups, such as the Buddhist Daisaku Ikeda approach and its theory of oneness of life, thus avoiding a false dichotomy between humans and nature.[15]

These ideas have been debated for some time and other terminology exists that means much the same thing, including *existence rights, ecosystem rights* and the *rights of nature.* The emphasis of all these is as with biodiversity rights less on the survival of the *individual* as the maintenance of *species, ecosystems* and *evolutionary potential.*

Is the concept utopian? Absolutely. No one expects to achieve this a hundred per cent, or anything like. Not only is it practically impossible but, as we'll discuss below, biodiversity rights will clash with many other factors, including questions of human rights and animal rights let alone development priorities and many less savoury pressures like conflict and criminality. But this doesn't make the ideas irrelevant or unimportant. The fact that human rights are abused on a daily basis throughout the world does not make a clear understanding of and commitment to human rights less important. Murder, sexual assault, slavery, homophobia, gender violence, religious tolerance, war … none of the world's moral commitments are fully achieved. But the fact that there has been a commitment helps move the debate forward and provides a moral framework against which to measure progress and impose sanctions.

A more fundamental objection is that a focus on preserving what currently exists is simply the wrong strategy given where the world is at present. There is a growing body of thought that conservation is manoeuvring itself into a corner by focusing too much effort on trying to preserve obscure, endemic species with limited distribution and probably limited impact on the global ecosystem and in trying to preserve "intact" ecosystems from change. This perspective argues that rather than a doomed attempt to halt ongoing extinctions we should focus instead on the new assemblages of organisms that are emerging as a result of our activities.[16] In this worldview, far from ending up with an impoverished planet, mixing so many species in new conditions will result in rapid evolutionary change and equivalent or greater diversity, better suited to the new conditions emerging during the twenty-first century – the "new wild".[17] The arguments are important; they're taking a bold look at the future and suggesting some positive alternatives. I agree that the whole concept of "natural ecosystem" needs a reboot.[18] But their conclusions don't work for me for a number of reasons. Some I'll explore later but in this context it is tantamount to washing our hands of any ethical responsibilities for what we've created in the first place and standing by while unique species drift into extinction because of our own actions. It is also unnecessarily defeatist. We can do better than this. Conservation efforts have already halted and reversed the decline of many species.[19,20] We certainly won't be completely successful but that is not reason to avoid even trying.

As noted above, one obvious drawback with the proposed definition of biodiversity rights is that it is by now practically unattainable for another reason; we have no way of being sure how much influence we are having on some species, with climate change alone making it almost impossible to know what "natural span in a functioning ecosystem" means any more. But in practice that doesn't really matter because we also know that the expected "background" rate of extinction is very low. Thus, we can interpret the implementation of biodiversity rights from the perspective of the timescale of human history as protecting as many species and ecosystems as we can. In turn, this implies maintaining evolutionary potential. This may seem like a lengthy process of navel-gazing to get back to where we started, but I'd argue that it is important precisely because these wider ethical concerns are often swamped and forgotten amongst more utilitarian arguments. The continuing confusion about what nature conservation means, even within nature conservation organisations, suggests we need to think more rigorously about these issues. This is not to say that the benefits that humans get from nature are unimportant.[21] But these are a practical rather than ethical consideration, and if they are the *only* consideration then a logical result is that if a better alternative comes along to replace an ecosystem service, the particular species or ecosystem involved ceases to have any importance. And when we start looking at how biodiversity rights interact with other rights – such as human rights and the rights of countries to develop – things start getting much more complicated.

But before moving on to discuss some of the challenges, I want to look in a little bit more depth at how rights relate to plants, fungi and ecosystems, because these issues are generally less well understood.

DO PLANTS AND FUNGI HAVE RIGHTS?

There is a genus of unicellular green algae called *Chlamydomonas*, about 150 species, found in freshwater and damp soil, gaining its energy by photosynthesis and reproducing sexually and asexually. *Chlamydomonas* also generally has two flagella, whip-like appendages which allows it to swim. The cilia of Chlamydomonas are very similar to those in mammals, making it an excellent vehicle for studying ciliary disease. Its ability to bridge the plant and animal lineages provides a powerful genetic and genomic platform and the genus has been the subject of a focused research effort for many years. Bottom line, they're really not very different from some unicellular animals, except they have a chloroplast and access energy directly from the sun. The extent to which they think things through probably isn't much different either. Should they be treated as fundamentally different?

The fact that plants and fungi should feature equally in biodiversity rights has already been asserted. It doesn't necessarily mean that animals and plants are always treated in the same way. As far as we know neither plants nor fungi "think" as animals do, nor to our best knowledge do they have anything like pain sensors, although they can often react very quickly to threats and pressures. However, and significantly, they demonstrate in abundance two of the other important factors linked to biodiversity rights: responses that suggest a drive to live and ensure generational survival and complex interactions between individuals, both of the same species and with other species.

In the 1970s, there were several books that suggested plants "thought", but in a way that we couldn't understand, responding positively to music and communicating via extra-sensory perception.[22,23] A bit like J.R.R. Tolkien's tree-herding ents. Building on the work of early plant ecologists, what scientists have been discovering since is perhaps even more interesting. Researchers have long been aware of different kinds of reasoning amongst colonial animals like ants, bees and termites, which display a "hive mind" capable of a collective, complex logic but is absent from individual members, apart from queens, if separated from the group. In other words, there are members of the animal kingdom that have ways of "thinking" that are extraordinarily different from our own. Star Trek makes use of this with the Borg, a hive mind based on the biology of colonial insects. The picture emerging of the way that plants interact with their environment is as complex and as partially understood as that of an ant colony. My old professor Phillip Wareing was one of the pioneers who explained how plant hormones interacted in directing the reactions of an individual plant.[24] Today we know that this is just the tip of a large and complex set of interactions. Plants communicate with each other, chemically and physically, both competing and mutually supporting, and with the birds, bats and the insects that pollinate them. They sense subtle changes in physical conditions, adjust their lifestyle accordingly and can often react quickly to threats and opportunities.[25] Through their seeds, which in some cases can remain viable for a hundred years or more[26] (potentially far longer in some circumstances), they demonstrate a prioritisation of species' survival alongside individual survival.

And the more we learn, the more complicated these reactions turn out to be. A sampling from Merlin Sheldrake's fascinating study of fungi includes descriptions of fungal networks extending over many square kilometres and living for thousands of years, slime moulds finding their way through complex mazes, fungal mycelium integrating multiple data streams to work out the optimal direction in which to grow, fungi hunting nematodes and fungi controlling the behaviour of ants to support their reproductive cycle. Trees and other plants are now known to communicate through fungal networks, exchanging nutrients between fungi and plants but also between different plants, even different species of plants, in effect operating sophisticated trading strategies.[27] No one is suggesting that they "think" in the same way that humans or other animals do. But nor can we any longer write them off as a simple toadstool.

Published back in 1972, Christopher Stone's book[28] *Should Trees Have Standing?* argued, amongst many other things, that trees and other plants should have rights of their own. He was writing at a time when we knew far less about the complexity of plants' interactions with other plants and with the environment, the case is even stronger today.

These rights don't translate in exactly the same way for plants and animals, any more than they do between humans and other animals. There was a tongue in cheek article in *The Guardian* newspaper recently asking if we shouldn't be confining our diet to nuts, fruits, leaves and other plant parts that can be eaten without damaging the whole organism.[29] Although some fruitarians do take this approach they are likely to remain a minority. The particular focus that conservationists often give to old trees is more due to their ecological and cultural values than any greater intrinsic value that an old tree has over a sapling. But the idea that plant *species* have the right to be given a fair chance of surviving makes perfect sense, although it creates a few tensions with some of the other associated rights movements, as we will see in the next chapter.

DO ECOSYSTEMS HAVE RIGHTS?

Conservation organisations have sometimes been slow in recognising the sophistication with which civil society – or at least those people likely to be interested in conservation – will respond to new concepts. The decision to set up a save the whale campaign split the management of Greenpeace,[30] which had previously focused only on nuclear weapons, yet whales proved of huge interest to vast sectors of the public. As mentioned before, Friends of the Earth and WWF were nervous about focusing on tropical forests – hot smelly places that would have little interest to their members – only to find themselves running some of their most popular campaigns. Most of the people who supported these efforts will never see a live whale, or walk through a tropical forest, but they instinctively saw that the forest as an ecosystem had the same right of existence as a species like a blue whale. At least I'm making that assumption; the early rainforest campaigns came at a time when the societal value of these places – for carbon, water services, etc. – was largely unrecognised. I was working for Friends of the Earth in Birmingham at the time and the reactions from people at talks, demonstrations and radio interviews was rooted more in an emotional recognition that such places had value rather than a utilitarian or economic rationale. That only came later.

Recognising that ecosystems have rights is important, partly because it is only within an ecosystem that species can fully express themselves and live sustainable, multi-generational lives that maintain evolutionary potential, and partly for the very practical reason that scientists have still only recognised and described a fraction of the species that exist on the planet. There is still an ongoing debate about the number of species existing and the number remaining to be discovered.[31]

But deciding on the parameters of ecosystem rights is more complicated. If we apply the same theory – that rights apply to an ecosystem as a whole rather than to individual examples of that ecosystem – then we need to categorise ecosystems. Whilst defining the boundaries of a species continues to be difficult, it is as nothing compared with defining the boundaries of an ecosystem. How different does an ecosystem need to be before it is considered unique and therefore have some type of collective rights? When does a habitat, such as a lake, or mire, or area of woodland, become an ecosystem? Global attempts to define sets of ecosystems, such as the biogeographical provinces,[32] global ecoregions,[33,34,35] biodiversity hotspots,[36] red list of ecosystems[37] and the global ecosystem typology developed by IUCN in 2022,[38] all encompass multiple different sub-ecosystems and habitats within them. Ecosystems are constantly changing and evolving; whole new ecosystems are likely to emerge as climate change bites, driving a process of evolution that has continued since the first simple lifeforms came into existence. Every single example of an ecosystem is unique and in theory has rights of existence, but this clearly creates an impossible situation. It also goes against the general principles that we are talking about here.

In practical terms conservation generally aims towards ecological representation, using broad ecosystem classification and staying alert for unique and important plant and animal associations that are hitherto unrecognised or undescribed. This sounds messy, but ecologists now have a fairly clear idea of the broad ecosystem and habitat types over most of the planet and can make informed suggestions about priorities. Prioritisation tools exist both in the form of global analyses, such as key biodiversity areas,[39] and approaches to look in detail at a particular area through systematic conservation planning; the two can also be used in combination.[40]

At the same time, because ecosystems are so complex, setting aside small areas doesn't guarantee survival except perhaps as an example of a representative ecosystem (in the way that natural World Heritage sites are supposed to be single examples of a range of ecosystems). So exactly how we put ecosystem rights into practice will inevitably be approximate but will draw on some combination of ecological restoration and endemism and at overall aims for ecosystem retention.[41] The significant international targets for retention and restoration of natural and semi-natural ecosystems agreed in 2022 show that ideas about conservation at scale are being increasingly recognised.

NOTES

1 Attfield, R. 2018. *Environmental Ethics: A Very Short Introduction*. Oxford University Press, Oxford.

2 See for instance Elliott, R. (ed.) 1995. *Environmental Ethics*. Oxford University Press, Oxford.

3 McShane, K. 2017. Is biodiversity intrinsically valuable? (And what might that mean?). In: Garson, J. Plutynski, A. and Sarkar, S. (eds.) 2017. *The Routledge Handbook of the Philosophy of Biodiversity.* Routledge, London: 155–167.

4 Baird Callicott, J. 2017. What good is it anyway? In: Garson, J. Plutynski, A. and Sarkar, S. (eds.) 2017. *The Routledge Handbook of the Philosophy of Biodiversity.* Routledge, London: 168–182.

5 Brosnan, S.F. 2006. Nonhuman species' reactions to inequity and their implications for fairness. *Social Justice Research* **19** (2): 153–185.

6 Orwell, G. 1945. *Animal Farm.* Martin Secker and Warburg, London.

7 Challenger, M. 2011. *On Extinction: How We became Estranged from Nature.* Granta, London.

8 Ripple, W.J., Abernethy, K., Betts, M.G., Chapron, G., Dirzo, R. et al. 2016. Bushmeat hunting and extinction risk to the world's mammals. *Royal Society Open Science* **3**: 160498.

9 Pearson, R.G., Stanton, J.C., Shoemaker, K.T., Ajello-Lammens, M.E., Ersts, P.J. et al. 2014. Life history and spatial traits predict extinction risk due to climate change. *Nature Climate Change* **4**: 217–221.

10 Watson, J. 2016. Bringing climate change back from the future. *Nature* **534**: 437.

11 Hu, H. and Jiang, Z. 2002. Trial release of Père David's deer *Elaphurus davidianus* in the Dafeng Reserve, China. *Oryx* **36** (2): 196–199.

12 Isbell, F., Craven, D., Connolly, J., Loreau, M., Schmid, B. et al. 2015. Biodiversity increases the resistance of ecosystem productivity to climate extremes. *Nature* **526**: 574–577.

13 Perrins, C. 1979. *British Tits.* New Naturalist Series 62. Collins, London.

14 MacDonald, B. 2019. *Rebirding: Restoring Britain's Wildlife.* Pelagic Publishing, Exeter.

15 Patterson, B. 2006. Ethics for wildlife conservation: Overcoming the human-nature dualism. *Bioscience* **56** (2): 144–150.

16 Meyer, S.T. 2006. *The End of the Wild.* MIT Press, Cambridge.

17 Pearce, F. 2015. *The New Wild: Why Invasive Species Will be Nature's Salvation.* Icon Books, London.

18 Dudley, N. 2011. *Authenticity.* Earthscan Books, London.

19 Mountfort, G. 1978. *Back from the Brink: Successes in Wildlife Conservation.* Hutchinson, London.

20 Balmford, A. 2012. *Wild Hope: On the Frontlines of Conservation Success.* Chicago University Press, London.

21 Stolton, S. and Dudley, N. (eds.) 2010. *Arguments for Protected Areas.* Earthscan, London.

22 Bird, C. and Tompkins, P. 1973. *The Secret Life of Plants.* Penguin Books, Middlesex.

23 Retallack, D. 1973. *The Sound of Music and Plants.* DeVorss and Co, Santa Monica, CA.

24 Wareing, P. 1970. *Growth and Differentiation in Flowering Plants.* Pergamon Press, Oxford.

25 Simard, S. 2021. *Finding the Mother Tree: Uncovering the Wisdom and Intelligence of the Forest.* Penguin Books, London.

26 Solberg, S.O., Yndgaardd, F., Andreasen, C., von Bothmer, R., Loskutov, I.G. and Asdal, Å. 2020. Long-term storage and longevity of orthodox seeds: A systematic review. *Frontiers of Plant Science.* DOI: 10.3389/fpls.2020.01007

27 Sheldrake, M., 2021. *Entangled Life: How Fungi make our World, Change our Minds and Shape Our Futures.* Vintage, London.

28 Stone, C.D. 2010. *Should Trees Have Standing: Law, Morality and the Environment,* 3rd edition. Oxford University Press, Oxford.

29 Beddington, E. 16th January 2023. If plants are so intelligent, should we stop eating them? *The Guardian.*

30 Hunter, R. 1980. *The Greenpeace Chronicle.* Picador, London.

31 Costello, M.J., May, R.M. and Stork, N.E. 2013. Response to comments on "Can we name Earth's species before they go extinct?". *Science* **341** (6143): 237.

32 Udvardy, M.D.F. 1975. *A Classification of the Biogeographical Provinces of the World.* IUCN Occasional Paper number 16. IUCN, Morges.

33 Olson, D.M., Dinerstein, E., Wikramanayake, E.D., Burgess, N.D., Powell, G.V.N. et al. 2001. Terrestrial ecoregions of the world: A new map of life on Earth. *Bioscience* **51** (11): 933–938.

34 Abell, R., Thieme, M.L., Revenga, C., Bryer, M., Kottelat, M., et al. 2008. Freshwater ecoregions of the world: A new map of biogeographic units for freshwater biodiversity conservation. *Bioscience* **58** (5): 403–414.

35 Spalding, M.D., Fox, H.E., Allen, G.R., Davidson, N., Ferdaña, Z.A. et al. 2007. Marine ecoregions of the world: A bioregionalization of coastal and shelf areas. *Bioscience* **57** (7): 573–583.

36 Brooks, T.M., Mittermeier, R.A., Mittermeier, C.G., da Fonseca, G.A., Rylands, A.B., et al. 2002. Habitat loss and extinction in the hotspots of biodiversity. *Conservation Biology* **16** (4): 909–923.

37 Rodríguez, J.P., Rodríguez-Clark, K.M., Baillie, J.E.M., Ash, N., Benson, J., et al. 2011. Establishing IUCN Red List criteria for threatened ecosystems. *Conservation Biology* **25** (1): 21–29.

38 Keith, D.A., Ferrer-Paris, J.R., Nicholson, E., Bishop, M., Polidoro, B.A., et al. 2022. A function-based typology for Earth's ecosystems. **Nature 610**: 513–518.

39 IUCN. 2016. *A Global Standard for the Identification of Key Biodiversity Areas.* Version 1.0. IUCN, Gland.

40 Smith, R.J., Bennun, L., Brooks, T.M., Butchart, S.M., Cuttelod, A., et al. 2018. Synergies between key biodiversity areas and systematic conservation planning approaches. *Conservation Letters* **12** (1): e12625. DOI:10.1111/conl.12625.

41 Simmonds, J.S., Suarez-Castro, A.F., Reside, A.E., Watson, J.E.M., Allan, J.R. et al. 2023. Retaining natural vegetation to safeguard biodiversity and humanity. *Conservation Biology* **37** (3): e14040. DOI:10.1111/cobi.14040.

5

WHAT ARE THE UTILITARIAN ARGUMENTS FOR PROTECTING BIODIVERSITY?

We're on the back of a pickup in the rainforest of Costa Rica and the skies have opened. It must be around the turn of the century. I'm so wet that it's almost funny, bouncing along a dirt track with the sounds of birds calling above the noise of the rain. Finally, we arrive at a hydroelectric station, quite small, generating three megawatts if I remember right and supplying some local townships. The owner is there to greet us and explain the set-up. I was involved in some of the new generation of hydro-projects in Wales and his system looks pretty sophisticated compared to what we were using back in the late 1970s. We're here because his power plant is an example of a Payment for Ecosystem Services (PES) scheme, which pays the custodians of a natural ecosystem for the benefits gained from that ecosystem, thus providing an incentive to keep it in good condition. In this case, he's paying a local community a few thousand dollars a year to maintain the rainforest around the site, to ensure that the water keeps flowing; mountain cloud forests tend to "scavenge" water from the air and increase net flow.[1] Back then, there was still an active debate about whether cloud forest really maintains water flow, something that is generally accepted now. Someone asked him about that. He said he knew about the debate but it was irrelevant; the amount of money that the local community wanted for management was low enough that it made perfect sense as an insurance policy. Let the scientists argue... at field level many people understand and benefit from ecosystem services or "nature's benefits".

We've spent the best part of 20 years researching, measuring and advocating for greater recognition of the ecosystem services from protected areas and other natural ecosystems. We've read the literature, talked

DOI: 10.4324/9780429346675-5

to researchers, developed and applied approaches for collecting information from local communities, looked at policy options and written more reports and books about these issues than I care to remember.[2] And we've been in good company, many people are doing the same, and finding ways to measure these benefits in ways that politicians and industry understand.[3] Most recently, the Intergovernmental Science-Policy Platform on biodiversity and Ecosystem Services has issued a major review of the diverse ways in which multiple values of nature are conceptualised and valued.[4]

Yet progress is slow; there are a multitude of good examples but it would be over-optimistic to say that the issue had as yet really gone to scale. Processes like PES and to an even greater extent the much-vaunted funding opportunities associated with carbon storage in natural ecosystems are still not having major impacts on a global or even national level.

Nonetheless, the intellectual arguments have been won and are increasingly reflected in national and international policy. Although the main focus here is on moral and ethical dimensions of species and ecosystems, it is worth summarising quickly what some of the practical benefits are as well. The following list is incomplete but gives a flavour of what we lose if we degrade or destroy natural ecosystems. Functioning, healthy natural ecosystems can supply all of the following and more.

Disaster mitigation: Economic losses from climate catastrophes have increased ten-fold over 50 years. Healthy natural ecosystems can help mitigate all but the largest natural disasters from:

- **Floods**: providing space for floodwaters to disperse and absorbing impacts with natural vegetation, thus slowing the rate of flow
- **Landslides**: stabilising soil and snow to stop slippage and slowing movement once a slip is underway, which can also help reduce loss of life in the aftermath of earthquakes
- **Tsunamis, typhoons and storms**: blocking water with coral reefs, barrier islands, mangroves, dunes and marshes
- **Drought and desertification**: ensuring sustainable rates of grazing to maintain vegetation cover and reduce soil erosion and dust storms pressure and maintaining drought-resistant plants for food.[5,6]

Water: water security is supported by forests and wetlands, both because they supply purer water than managed catchments and in some cases they also increase net flow:

- Well-managed natural forests and wetlands provide purer water with less sediment and pollution than water flowing through other catchments
- Some natural forests (especially tropical montane cloud forests) and vegetation associations like the Andean *paramos* also increase total water flow
- 33 of the world's 105 largest cities use protected areas as sources for a significant amount of drinking water; ten more manage forests for clean water.[7]

Biodiversity for Food and Agriculture is a vital and under-valued contribution to food security:

- Estimates of the global value of crop wild relatives (CWR) vary from hundreds of millions to tens of billions of US dollars per year[8]
- Natural ecosystems also supply pollinators, predators of pests and soil animals that maintain fertility
- A total of 250 million people are dependent on small-scale fisheries for protein and fisherfolk often support protection of marine and freshwater reserves, which increase both size and populations of fish and spillover replenishes surrounding fishing area.[9]

Faiths and cultural values: ranging from sacred sites to aesthetic or historical values:

- All major and many minor faiths recognise sacred natural sites (SNS), even Islam and Christianity in places[10]
- Sacred values (SNS, pilgrimage routes and sacred buildings) occur in many protected areas and many sacred sites outside protected areas also have high biodiversity values and these often have rich biodiversity[11]
- Cultural values range from very local values associated with a particular view, tree or natural feature to globally significant landscapes that attract visitors from around the world.

Health: natural ecosystems supply medicinal plants, provide space for physical exercise and mental relaxation and management can help to control spread of disease:

- Health benefits can come from habitat protection – e.g., control of zoonotic diseases

- Many people (80 per cent in Africa) rely on traditional medicines and 60 per cent of medicinal plants are collected from the wild; protected areas protect many species from extinction
- Protected areas are used to promote exercise and to help mental health patients and people with social difficulties.[12]

Climate change: natural ecosystems are critically important for carbon sequestration and ecosystem-based adaptation to climate change. Many adaptation strategies are similar to those described above

- Sequestration and carbon storage is important in many ecosystems, particularly forests and peatlands but there are also huge stores in grassland, savannah and marine ecosystems
- Protected areas provide ideal delivery mechanisms for both sequestration and ecosystem-based adaptation.[13]

Note that these are very direct benefits, natural ecosystems also drive the weather cycle, soil formation, the photosynthesis on which all agriculture and human nutrition ultimately survives. Furthermore, research shows that ecosystem services tend to increase with diversity,[14] as does their resilience, the ability to withstand change. There are therefore abundant arguments to conserve biodiversity for quite selfish reasons, indeed failure to do so would genuinely spell disaster. Fortunately, the benefits listed quickly above already have many organisations and researchers championing their cause and I won't be discussing them further here.

NOTES

1 Hamilton, L. with contributions from Dudley, N., Greminger, G., Hassan, N., Lamb, D., Stolton, S. and Tognetti, S. 2008. *Forests and Water*. FAO Forestry Paper 155. Food and Agricultural Organization, Rome.

2 Stolton, S. and Dudley, N. (eds.) 2010. *Arguments for Protected Areas*. Earthscan, London.

3 Neugarten, R.A., Langhammer, P.F., Osipova, E., Bagstad, K.J., Bhabati, N., et al. 2018. *Tools for Measuring, Modelling, and Valuing Ecosystem Services: Guidance for Key Biodiversity Areas, Natural World Heritage Sites and Protected Areas*. IUCN, Gland.

4 IPBES. 2022. Summary for Policy Makers of the Methodological assessment report on the diverse values and valuation of nature of the intergovernmental science-policy platform on biodiversity and ecosystem services. IPBES Secretariat, Bonn.

5 Kumagai, Y., Furuta, N., Dudley, N., Naniwa, N. and Murti, R. 2013. Responding to disasters: The role of protected areas. *PARKS* **19** (2): 7–12.

6 Dudley, N. Buyck, C., Furuta, N., Pedrot, C., Bernard, F. and Sudmeier-Rieux, K. 2015. *Protected Areas as Tools for Disaster Risk Reduction: A Handbook for Practitioners.* IUCN and the Ministry of Environment, Gland and Tokyo.

7 Dudley, N. and Stolton, S. 2003. *Running Pure.* World Bank and WWF, Washington, DC and Gland.

8 Stolton, S., Maxted, N., Ford-Lloyd, B., Kell, S. and Dudley, N. 2006. *Food Stores.* WWF and the University of Birmingham, Gland and Birmingham.

9 Stolton, S. and Dudley, N. 2010. *Arguments for Protected Areas.* Earthscan, London.

10 Dudley, N., Higgins-Zogib, L. and Mansourian, S. 2005. *Beyond Belief.* WWF and ARC, Gland and Bath.

11 Dudley, N., Bhagwat, S., Higgins-Zogib, L., Lassen, B., Verschuuren, B. and Wild, R. 2010. Conservation of biodiversity in sacred natural sites in Asia and Africa: A review of the scientific literature. In: Verschuuren, B., Wild, R., McNeely, J. and Oviedo, G. (eds.) *Sacred Natural Sites: Conserving Nature and Culture.* Earthscan, London: 19–32.

12 Stolton, S. and Dudley, N. 2011. *Vital Sites.* WWF, Gland.

13 Dudley, N., Stolton, S., Belokurov, A., Krueger, L., Lopoukhine, N., et al. 2009. *Natural Solutions: Protected Areas Helping People Cope with Climate Change.* IUCN-WCPA, TNC, UNDP, WCS, the World Bank, and WWF, Gland, Washington, DC, and New York.

14 Díaz, S., Fargione, J., Stuart Chapin, F. III and Tilman, D. 2006. Biodiversity loss threatens human well-being. *PLoS Biology* **4** (8): e277. DOI:10.1371/journal. pbio.0040277.

6

RIGHTS IN CONFLICT

It will be clear from the whole tenor of this book that I think the concept of biodiversity rights is critically important in deciding how we should live our lives. The emerging recognition of such rights is one of the major philosophical steps forward in the late twentieth and early twenty-first century. Albeit imperfectly understood and for a long time un-named, biodiversity rights have shaped my personal and professional life since I first remember worrying about disappearing species sometime in primary school. And importantly, this is a concept that has emerged more or less organically from the bottom up, with civil society often running ahead of many conservation professionals in an instinctive understanding that the issue is biodiversity as a whole, rather than just individual species.

But it has taken much longer to recognise that biodiversity rights can also have some significant associated costs, at least in the way that they have been addressed until now. Any social change has costs to someone: the abolition of slavery hit the pockets of anyone involved in the trade, the changing role of women has made the jobs market harder for men, health campaigns against tobacco impact poor farmers in countries like Malawi and so on. But in the case of biodiversity rights, taking account of the needs of other species and ecosystems bangs up against some very fundamental ethical issues, including human rights and paradoxically also sometimes animal rights. The question of what is right or wrong in a particular issue can become very murky. It is also unfortunately fair to say that the way that the conservation movement has until recently approached things has not always helped.

J. Baird Callicott suggested, in a now classic essay, that environmental rights of the type discussed here form a mutually opposing equilateral

DOI: 10.4324/9780429346675-6

triangle with humanism (in this case particularly human rights) and sentientist rights, here broadly meaning animals rights.[1] He makes an important point, but there are ways of reconciling these issues – one of my purposes here – a stance the author shares himself. He has sometimes regretted writing the essay at all, but I think it sparked a really useful debate.[2] Suffice to say that in very practical terms reconciling biodiversity rights, animal rights and human rights has become one of the trickiest issues facing anyone involved in practical conservation at the present time. To some extent, this is inevitable because the three approaches do indeed have different priorities, but there are also important areas of common ground. Finding these and working out how the three can work together in harmony is one of the most urgent challenges facing the sustainable development movement today.

Complicated ethical questions are not confined to the way that biodiversity rights may clash with other rights but can also result in some internal tensions. Whose rights? Is all biodiversity created equal? That is the inference of our definition of biodiversity rights but put more subtly, do some species have a disproportionately larger role in the survival of other species and thus tacitly assume a greater importance overall? What about alien species, recently introduced to ecosystems by human actions. Conservationists often treat them as "the enemy", with severely curtailed rights but is this necessarily or always true? How do the various utilitarian values of species and ecosystems – their ecosystem services – impact on their rights, if at all?

In the following section, the core of the book, I examine some of these questions, both by reviewing how others have addressed these issues and by proposing some responses of my own.

WHAT ARE ANIMAL RIGHTS AND HOW DO THEY RELATE TO BIODIVERSITY RIGHTS?

We are in Kathmandu in 2001, sitting high on some temple steps and looking down on what proves to be a funeral. There is a river running nearby and a troop of rhesus macaque monkeys (*Macaca mulatta*). The monkeys are sacred to the Buddhists and roam freely around the temples in the city, tolerated even at a solemn occasion like this. As we watch, a group of the youngest monkeys begin a game, swinging on a vine out over the river and dropping in, swimming ashore and scrambling up to repeat, again and again. Take a look on social media and you'll see plenty of humans doing much the same. And as with ourselves, the squeals got louder and louder as excitement grew and the play gradually became more frenetic, till one of the older males wandered over, grabbed each of the youngsters by the scruff of the neck and gave them a firm slap on the backside, sending them scuttling off into the trees.

Which is by way of stating what I hope is the obvious, animals are sentient beings and, as we get nearer to our branch of the evolutionary tree, not as different from ourselves as we might like to think. To a greater or lesser degree many animals play, they show affection, care for their offspring, demonstrate loyalty, build deep emotional ties with other creatures, sometimes not even of their own species. They can reason to a certain extent, feel fear and pain, exhibit jealousy, the embarrassment of undergoing loss of face, grief at the death of a group member, and a wide range of other emotions. Researchers are also looking at the ability of species to make what might be termed aesthetic judgements.[3] The extent to which different species exhibit such traits varies enormously, but the general point holds.

Recognition of the moral standing of sentient beings has ebbed and flowed over time. Many ancient and traditional societies have understood the intrinsic value of other forms of life very clearly and many so-called more "advanced" people have not. Our oldest stories, myths, legends and folklore commonly treated animals as reasoning beings, a perspective that was much more common than is often recognised. But throughout written history it is fair to say that, until recently, the majority view at least in the West was that the world had been created for humans – or more specifically for 50% of humans, men – and that plants and the rest of the animal kingdom were there to serve his purpose. Aristotle thought so, as did the Stoics. Christians followed suit (to a greater extent than the Jews) with the great mystic St. Thomas Aquinas arguing that *all animals are naturally subject to man*.[4] Rene Descartes, the French philosopher, took this to a logical conclusion by arguing that animals are mere automata, capable of complex operations rather like a clock but without sensations.[5] Women, the poor and in time anyone with a different skin colour were all often categorised somewhere in between animals and Man; theologians had semi-serious debates about whether women possessed souls[6] and much more recently many evolutionary studies argued that women were intellectually and physically inferior.[7]

But there were exceptions. Minority groups, like the Cynics, Sceptics and Epicureans, had long argued that humans were not necessarily the centre of the universe. In the seventeenth and eighteenth centuries, many British clergymen were assuming that animals had souls and that they would be present in the Kingdom of Heaven. In France, the hard-line approach promoted by Descartes did not actually remain ascendent for very long.[8] And there had always been what historians refer to as "privileged species", such as dogs, cats and horses, which many people treated as honoured members of the family. Thinkers from as widely disparate starting points as Friedrich Nietzsche, Peter Kropotkin, Albert Schweitzer, John Stuart Mill, Jeremy Bentham and Mahatma Gandhi all argued the case for a more humane attitude towards animals.[9] In the nineteenth and early twentieth century, the women's movement often linked their cause to the rights of animals, particularly opposing vivisection.[10]

In the twentieth century the animal rights movement snowballed, opposing hunting, vivisection and animal testing, factory farming, meat eating and more. Philosophers like Mary Midgley[11] and Peter Singer[12] made the intellectual and moral case for animal rights cogently and persuasively. Singer argued that all beings capable of feeling pain should be treated equally, i.e., that this is a more important distinguishing feature than intelligence.

Some activists in the animal rights movement have also, and to the detriment of the overall message, become radicalised, with death threats against medical researchers using animals, releases of mink from fur farms and similar actions, some of which directly undermine biodiversity rights. More sedately, the number of vegetarians and vegans has been increasing steadily in many parts of the world and it is a rare café and restaurant in Western Europe that does not now have vegetarian and vegan options on the menu. Amongst consumers of meat and dairy products, growing numbers avoid factory-farmed animals, opting for free range eggs, and organic or other forms of meat and milk products that give some guarantee of compassionate treatment. These changes relate to individual animals and almost entirely to domesticated animals, many of which have been anthropomorphised in countless books, films and toys. Many children growing up in an urban environment do not connect the meat they eat with the farm animals that appear in their picture books and cartoons.

Popular confusion between the rights of *individual* animals to live their natural lifespan in natural conditions, and the rights of species to survive *as species*, is so profound that it sometimes distorts the whole debate about the ethics of conservation. It is important enough that it needs to be addressed right at the beginning of any debate about biodiversity rights.

There is in fact surprisingly little overlap and some identifiable philosophical and ethical conflicts. Tensions between the two philosophies have been recognised for some time. Human moralists argue that individual organisms, rather than species or ecosystems, should be the focus of our ethical concerns, with sentience – the ability to experience pain – as the characteristic needed for organisms to receive full ethical consideration.[13] This itself creates some questions, including how we know whether less expressive animal species are experiencing pain. There is an unresolved debate about whether fish feel pain for instance.[14] I'm not sure that a fish gradually expiring on the deck of a trawler has a much easier time of it than a cow quickly despatched with a captive bolt.

The conflict between animal rights and biodiversity rights is obvious and dramatically demonstrated by the actions of a proportion of extreme animal rights activists. "Liberating" non-native mink (*Neogale vision*) from fur farms has had disastrous impacts on the ecosystem[15]; in Britain feral mink populations created by accidental and deliberate releases have driven the once common water vole (*Arvicola amphibius*) virtually to extinction in many areas. Many other species have suffered as a result. The actions of a fringe should never be confused with the opinions of the majority and the animal rights issue has a rich and well thought through philosophical basis.[16] But the differences between the two philosophies are also true at a deeper, theoretical level. To date they have seldom been explored thoroughly enough to allow the development of a very meaningful dialogue, let alone resolution to these differences. Some of the strongest arguments against animal rights have come from within the wildlife conservation movement, with its long history of links with hunting, sustainable harvest and at least in some countries support for Indigenous and other traditional hunting communities.[17]

Animal rights philosophy – biocentrism – argues that it is arrogant and illogical for humans to separate themselves from the rest of the animal kingdom and that other species have similar moral rights and deserve similar protection against human-originated cruelty, incarceration, suffering and killing. *Meat is murder* as the slogan says. By nature of our greater capacity for reasoning, humans are clearly capable of aiming for higher standards in treatment of other species: the fact that a fox will indiscriminately slaughter all the hens in a henhouse does not give us the excuse to behave in the same way. In a thoughtful, broadly Kantian discussion the philosopher Tom Regan argues that non-human animals have moral rights, although he explicitly rejects Emmanuel Kant's original caveat that this only applies to rational beings. Regan calls for the total abolition of the use of animals in science, the total dissolution of the commercial animal agriculture system, and the total elimination of commercial and sport hunting and trapping.[18]

This is a fairly extreme position and there are myriad interpretations of animal rights. Decisions are made by individuals on a case-by-case basis, often relying more on emotion than logic. In practice, many people operate a kind of sliding scale, paying most attention to the rights of larger and more intelligent animals. Many vegetarians will swat a mosquito if it lands on their arm for example and a significant proportion of people will eat fish but not mammals or birds. Some philosophies and faiths, such as the Jains, afford all animal species equal rights, right down to normally despised species like flies and mosquitoes, but these are a minority.

Culture, tradition and religion all play an important role in determining what is considered ethically acceptable: many meat eaters in the west would reject the idea of eating a dog or a cat although there is no reason to think these species more intelligent than, say, a pig or a cow[19] and they are routinely eaten elsewhere. Many meat eaters in Britain will balk at consuming a horse[20] (and there was a recent scandal about illegal selling of horsemeat in the UK) but this is standard fare in France and Switzerland, which are not in other ways regarded as having different "ethical" cultures. The Vietnamese are one of several Asian countries where many people eat dogs. Whale meat, long considered a cheap substitute meat, is now completely taboo for many people but has a weird cachet for a minority. It is also a traditional meat for some Indigenous peoples, who regard campaigns against their diets to be a form of colonialism, coming from people who are happy to eat equally intelligent animals themselves.[21] In the brief period that I kept farm animals, it was the pigs that I found most intelligent and most likeable; an opinion I've since heard expressed by several farmers. In some countries the very fact that an animal is rare adds to its value as food; tiger meat now apparently costs the same on the black market as an equivalent weight in heroin. Status is important here; rich and powerful people have traditionally demonstrated their wealth by eating rare or otherwise difficult to obtain plants and animals.

More logically, many meat eaters extend clear ethical concerns about the lifestyles of the individuals they consume. An increasing number of people who eat meat and dairy products now look for evidence of treatment of farm animals that accords them certain rights, such as free-range conditions, an organic certificate, minimal transport and "humane" killing methods, which minimise fear and suffering when the animals are slaughtered. I know some people who will only eat meat if it has been killed in the wild, for similar reasons. Veal, paté de foie gras, two-week old lambs and other specialised meat products are avoided by many meat eaters and for example production of foie gras has been banned by several countries.

These different moral positions occur against the background of a huge change taking place over the past 200 years or so in what people consider acceptable treatment of animals[22] (in line with similar changes in attitudes towards what is acceptable treatment of humans). Sports like badger baiting and cockfighting have been driven underground in most societies, day-to-day cruelty is highly censured and animal charities attract high levels of support. Evidence of people being cruel to animals creates a huge public outcry in many European countries and social media shaming, something that would have been undreamed of a century ago and would still be unlikely in many parts of the world. There is no reason to suppose that this shift in moral compass has completed its course.

Virtually everybody who supports biodiversity rights also supports some aspects of animal rights, but the details of peoples' ethical choices vary dramatically. Direct conflicts between animal rights and biodiversity rights emerge when individual animals' rights clash with the function (the "rights") of whole ecosystems or with the existence of other species: either plants or those animals implicitly judged to be less "important". Some of these conflicts emerge because of misunderstandings or sentimentality: we can assume that the activists who release mink from fur farms are not deliberately intending to create a mass slaughter of smaller animals. Passionate resistance to eradicating introduced hedgehogs from a small Scottish island led to death threats against Scottish civil servants but also endangered whole populations of nationally important ground-nesting seabirds, whose nests were being raided by the hedgehogs. The UK's cat population, almost by definition owned by animal lovers, kills an estimated 300 million animals a year, mainly songbirds[23]; impacts in countries like Australia have been even more devastating. Like many actions driven by strong emotions, logic sometimes gets scrambled, making the resulting controversies particularly difficult to resolve.

Other conflicts occur at a different level. Someone interested in biodiversity rights would be likely to support culling a population of elephants in a savannah if, in the absence of effective predation and where suitable habitat is mainly restricted to protected areas, their numbers rose to levels where they are destroying whole woodland habitats and associated species. Several national parks in Africa have faced that dilemma.[24] When elephant numbers become so high that they are causing serious threats most conservationists prefer to kill a whole family group rather than spreading the social impacts among many groups. In these cases, the need to restrict a population has usually come about because of human land-use change and removal or elimination of predators. But for someone interested in animal rights the prospect of killing highly intelligent beings, with a complex family structure, loyalties and emotions, for the sake of, say, vegetation and invertebrates is far more ethically ambivalent.

This dilemma was played out in a situation almost designed to test the meeting of the two philosophies: on the Californian island of Santa Barbara, introduced hares, cats, sheep and particularly rabbits were destroying the habitat of a small succulent plant found no-where else in the world, threatening its extinction.[25] Following a philosophy of biodiversity rights there is no question that the rabbits should be eliminated to protect a unique species. But from an animal rights perspective killing several hundred sentient beings for the sake of a plant that is irrelevant from an economic perspective and unlikely to have much evolutionary influence due to its isolation makes no sense at all. In fact, all the invasive species were removed and the plant recovered its numbers.[26]

The debate about so-called "compassionate conservation"[27] shows up some of the problems. Compassionate conservation argues for an approach that pays greater attention to the rights of sentient animals and focuses on issues relating to "do no harm", the importance of individual animals and the need for peaceful co-existence between animals and humans. Critics argue that there is too much focus on charismatic species and a reluctance in particular to control (kill) alien predators, which will lead to the death of larger numbers of native species.[28] Analysis in 2021 found criticisms centred on ethical foundations, scientific credibility, clarity of application and subsequent threats to biodiversity and conservation.[29] What the debate also shows, quite starkly, is the distance between the two schools of thought.

These differences are obvious in the attitudes of people working for conservation organisations. Bodies like WWF probably have more vegetarians on staff than the population average, but in my experience not to any significant extent and a proportion of these will be motivated by consideration of food security issues, health concerns or the links between animal raising and climate change rather than ethical considerations for other species or the conditions those species are raised in. And at the same time conservation organisations often include a fair number of hunters or anglers, especially people hunting for food. I have been surprised over the years about how seldom these issues are even discussed, as if colleagues have an unspoken agreement to treat ethical differences in attitudes to animals in the same way as differences in religious faith: something to be acknowledged tacitly but not debated. But contempt for extreme animal rights activists, like those releasing farmed mink into the general environment, is almost universal amongst conservationists; ironically so as many people in society at large continue to assume the two are identical.

There are clear dangers in taking the distinction between species and individuals to an extreme. The twentieth century showed with horrible detail what can happen when political philosophies focus entirely on concepts like "humanity" or "mankind" and forget about the individual: the gulags and concentration camps, starvation in the Soviet Union and the Great Famine of China were all justified by their perpetrators in terms of some theoretical overall good. If we focus only on the species and ignore the individual we risk a similar distortion. Therefore, I am not arguing that we should forget about animal rights in favour of a grander concern for the entire planetary biosphere, but that instead we need to engage in active debate about how these two worthy concepts can be integrated, where trade-offs are possible and what compromises need to be made. When, if ever, do the rights of species trump the rights of individuals? Do "invasive species" somehow have less "rights" than the native species that they are outcompeting, and if so why? Do animals always have more rights than plants, or is this attitude simply anthropomorphism? Does the life of an ancient pine tree have any less moral standing or intrinsic value than the life of the caribou that grazes underneath its canopy? Do rights increase with intelligence? If so, animal rights always trump those of plants, at least until we get down to the size of unicellular plants and animals where the difference probably isn't really that great.

Note that Tom Regan's call to eliminate animals from medical research, agriculture and sport and commercial hunting says nothing against killing animals to protect other animals, or plants. The culling of rabbits on Santa Barbara Island, or goats on the Galapagos, rats on South Georgia, hedgehogs in the Scottish Hebrides and the daily trapping of stoats and weasels to protect ground nesting birds in New Zealand would not be covered by his ban. Practically, as we have seen, many animal rights activists strongly oppose such actions, sometimes with threats of violence. If animals are really to be judged as equivalent to humans, does that mean that killing of humans is justified to save another species? This is something that most of us would instinctively reject. If we do so, we are assuming that humans are worth "more" than other animals. This stance would not be at odds with the opinions of philosophers like Peter Singer but how far does this go? If humans really are more important than animals, does this mean that human needs and wants are always more important than other animal needs? Or plant needs? Deciding on the point at which the rights of other species outweigh the rights of humans is fraught with dangers which we'll look at in the next section.

BIODIVERSITY RIGHTS AND HUMAN RIGHTS

A much more complex issue relates to how biodiversity rights stack up against human rights. Do other species ever have rights over humans? What rights and when? Again, the situation is complicated and even more sensitive.

Many, probably most, actions to conserve other species will impact to the detriment of at least some people. A ban on hunting a rare group of monkeys will not usually be welcomed by the people who hunt them. Local fisherfolk cooperating with conservation agencies to set up zones restricted to artisanal fishing will be resented by offshore fishing fleets from further afield. Talk of win-win solutions is usually naïve and compromises need to be negotiated.

I asked above if the lives of other species would ever be prioritised over human lives. On one level we do this all the time; any action to conserve or reintroduce a species that can attack or even kill a human is making a judgement call about levels of acceptable risk. Otherwise, we would be doing what our ancestors did and hunting any potentially dangerous species to extinction, as has happened with bears in Britain, tigers in Singapore, lions in many parts of Africa and wolves in large areas of North America. Apart from specimen groups in zoos and fenced game parks we would say good-bye to tigers, lions, jaguars, elephants, rhinoceros, hippopotamus and crocodiles, all species that we could realistically plan to wipe out if we put our minds to it, and we'd be waging all-out war against any poisonous snakes and sharks. The fact that we do not, and conversely we actively protect the species listed above, shows that someone is making a conscious or unconscious judgement call about levels of risk. As we do in many other areas of human society, such as setting speed limits on roads, pricing of pharmaceutical products or the stringency of controls on alcohol and tobacco. And as with these other examples, our judgements about acceptable risk from animals is not always strictly logical.

Two issues are particular important with respect to conservation: tensions relating to human-wildlife conflict and threats from dangerous animals on the one hand and forcible removal of people to make space for nature – particularly national parks and wilderness reserves – on the other.

These are not the only issues by any means but they are the ones that cause the most controversy and about which many professionals need to make decisions on a daily basis.

Coexistence with dangerous or other "problem" animals

Tigers declined by an astonishing 97 per cent during the twentieth century, with whole countries losing their populations altogether. Tigers faced a barrage of threats: habitat loss, loss of prey species, and increasing levels of persecution, latterly linked with a market in tiger products for traditional medicines and as a high-status source of meat. People started talking seriously about extinction in the wild. The 2010 "Tiger Summit" in St Petersburg aimed to stop the decline and to double tiger populations over the following decade. The ten years following saw a gradual turnaround and, at least in some countries, the target appears to have been achieved. (Tigers are difficult to count; it may be that some of the apparent increase is due to greater monitoring effort, but there has also clearly been a genuine increase.) Although the survival of wild tigers remains fragile, and numbers are still falling alarmingly in several South-East Asian countries,[30] evidence of an upward global population trend was welcome news with which to start the UN Decade on Ecosystem Restoration. Idealistic strategies made in 2010[31] now seem justified.

But is it really good news? Probably not for everyone. I'm writing in Wales, a country where I can be reasonably assured of walking around without danger from animals, unless the sleepy old bull in the hill above my house suddenly turns nasty. Enthusiasm about tiger survival is easy for those of us who live a long way from the tiger range; for whom seeing wild tigers is at most a rare and exciting treat, peering through the window of a secure safari truck. People who co-exist with tigers daily may not be so enthusiastic about their recovery. Tigers are the world's largest and fiercest cats; particularly when under pressure for food and resources they will take livestock and they pose a constant threat to humans; people are killed by tigers every year. Tigers are sensitive to human disturbance and ideally need large areas protected for their conservation, but this inevitably leads to conflict with people.[32] In India for instance, 320 people were killed by tigers between 2014 and 2020,[33] in Nepal 20 people were killed from 2010 to 2014.[34] Human deaths are still a rare occurrence in Bhutan,[35] where populations of both people and tigers are low. But it is not just the human tragedies that are an issue. Between 400 and 600 cattle and goats are killed by tigers and leopards each year around Kanha Tiger Reserve, central India,[36] with similar or higher losses in Corbett National Park, further north.[37] Loss of a cow can be a devastating economic blow to a poor smallholder family and compensation schemes are at best often uncertain and slow. Plans to increase tiger populations in densely populated areas of Nepal and India, and proposals to reintroduce tigers in countries such as Kazakhstan[38] and Cambodia[39] where they have been extirpated, will be controversial. When I worked on an assessment of South Korea's national parks[40] there were impressive plans to reintroduce several species, but not the tiger even though it is Korea's "national animal". When conservationists implement plans to increase tiger numbers they will be inadvertently creating the means whereby other people will be killed. Furthermore, the deaths will be overwhelmingly amongst the poorest members of society who cannot afford to stay out of danger areas because they need to collect firewood, livestock fodder or plants to eat. For the last decade we have been working on improving standards of tiger conservation,[41] we're directly implicated. And although I'm talking about tigers here, similar or greater death tolls are associated with lions, jaguars, leopards, and with hippopotami, elephants, crocodiles and other dangerous animals such as snakes and fish such as sharks. Interactions are also changing over time, as human populations press further into hitherto unpopulated areas. After a long period when bears in Europe have generally stayed away from humans, they have again become dangerous in Romania, for reasons that are complex but have an immediate impact on human use of forests. There were signs warning us to keep out of forests in the afternoon when we visited Romania in 2008, where until recently bear attacks had been a very rare occurrence.

This isn't a unique conundrum of course, any decision to straighten a road, reduce medical funding, sell tobacco or advertise unhealthy foods will cause deaths. But the fact that society accepts these costs doesn't mean that they are trivial. Conservation of large predatory mammals is a very clear way in which human rights and animal rights can come into conflict.

Clear but not simple. People living in the tiger range often have a much more philosophical approach to the dangers than those who have grown up in a more secure environment. Religious faith plays an important part in this. Tigers are closely linked with Taoism,[42] Buddhism[43,44] and in Hindu traditions.[45] In Bengal, the tiger god was worshipped by both Hindus and Muslims.[46] Many tiger reserves contain important religious buildings, such as Periyar Tiger Reserve in India, visited by 10 million devotees every year.[47] Further to the east, the traditional Tungusic, Udege and Nanai peoples of Siberia consider the tiger a near-deity.[48]

But does this translate into willingness to conserve?[49] The evidence is mixed and changing. Buddhist traditions have close links with conservation[50] and in Bhutan for instance the tiger is considered sacred[51] and conservation success is partly due to people's religious faith.[52] Beliefs about tigers are identified as major factors in determining tolerance levels in the Indian Sundarbans.[53] Attitudes also depend on where problems occur; villagers in Sumatra felt tigers should be killed if they attacked people near villages but not if attacks took place in the forest and there was also a belief that if tigers killed someone who had committed adultery, the cats were enforcing a moral code.[54] In January 2014, Indonesia's powerful Islam council of scholars announced the first ever biodiversity fatwa, requesting all Muslims to safeguard biodiversity and particularly threatened species like the tiger.[55] Countries aiming to restore tiger populations, such as China and Kazakhstan, hope to build public support based on the role of tigers in national cultures.[56] But we shouldn't overstate the case. Research on attitudes to snow leopards and wolves amongst Muslim and Buddhist communities in northern India found religious belief to be statistically insignificant in shaping opinions, although active Buddhists were more likely to be tolerant of carnivores.[57]

Coexistence is defined as a dynamic but sustainable state in which humans and wildlife co-adapt to living in shared landscapes, where human interactions with wildlife are governed by effective institutions that ensure long-term wildlife population persistence, social legitimacy and tolerable levels of risk.[58] Developing coexistence instead of conflict is identified as a key step in securing a future for large predatory or dangerous animals like tigers and elephants.

Co-existence relies on a complicated series of trade-offs,[59] and continues to carry risks for both people and dangerous animals. Human casualties are the most significant cost of co-existence,[60] although livestock predation is commoner and an important and understandable source of tension.[61] Co-existence tensions are also not limited to physically dangerous animals. Primates and deer have a huge impact on farms, home gardens and trees. Human wildlife conflict is one of the commonest issues that protected area managers are addressing on a day-to-day basis in many parts of the world. "Payments to encourage co-existence" can help, including compensation and insurance schemes, revenue sharing mechanisms and various forms of direct payments linked to conservation success.[62] But these can also perversely increase resentment, by framing the issue more in terms of conflict than co-existence, and looking at costs rather than the benefits of having large animals still roaming wild in the ecosystem.

Far from being a developing country issue, three of the top four wealthiest countries in the world in 2050 could well be within the tiger range, with a rapidly growing consumer class, strong interest in conservation and rural depopulation potentially creating more tiger habitat. I have been focusing mainly on tigers here partly because this is probably one of the most contentious issues and because we have looked in detail at the question of co-existence with tigers for WWF.[63] That project came up with a number of key recommendations, to:

1   Adopt impactful top-level goals on human-tiger coexistence and directly link such strategies to the UN Sustainable Development Goals and other international targets.
2   Rapidly expand direct community involvement in tiger conservation decision-making and in doing so, create new forums for direct dialogue with "tiger" communities at all governance levels (local, national, regional).
3   In consultation with local communities, design and implement new policies that reduce the costs and expand the benefits of living with tigers, both as a matter of fundamental fairness, and in recognising the critical role local peoples play in maintaining tiger populations.
4   Significantly increase investments for tiger conservation outside traditional protected areas systems, utilising social science expertise and facilitating the processes that can lead to the formation of new community conserved areas within tiger areas.

All these things are easy to write but challenging to implement.[64] Understanding the source of conflicts, long-term engagement with often disparate local communities and continual willingness to learn and adapt are all essential elements of success.

Around the world, communities are struggling with human-wildlife conflict,[65] from baboons raiding crops to rogue elephants trampling through villagers, river dolphins damaging fishing nets and crocodiles taking children when they collect water from streams and rivers. A combination of community empowerment and sophisticated technology can help reduce risks. Radio trackers and mobile phones allow villagers to know when problem animals approach crops or houses for instance.[66] An Information and Communication Technologies-based system in Botswana's Okavango Delta significantly reduced financial losses from livestock predation and was well received by local people. The system relies on GPS tracking technology and agreement of 'geofences', or virtual boundaries, with alerts triggered when a radio collared lion crosses them. Information was passed to the village headmen and herders when a lion was within 8.0 km linear distance of grazing land and within 5.0 km linear distance of village locations.[67] It is becoming easier to establish virtual, moveable "fences" which, if crossed, will result in a quick and unpleasant shock to an animal through something embedded in their skin. Electronic sound devices ("pingers") can scare river dolphins away from fishing nets. Solar-powered lighting systems (fox lights) deterred tigers from entering villages in India.[68] Electric fences[69] and stockades[70] are used to protect against predators throughout the world. Predator-proof night corrals for smaller domestic animals[71] are constructed to keep out predators.[72]

These tools and strategies are all important steps in reducing conflict and management tools are likely to become increasingly sophisticated in the future. Of course, purists will note that they also change the species from being wholly "wild" to animals that are in effect being managed. This kind of trade-off is likely to become increasingly common.

A broader response, a recognition that human wildlife conflict needs to be recognised and addressed, is an important element of management of biodiversity rights. Specialists increasingly see the word "conflict" as damaging in itself, setting up other species as the enemy, and prefer to refer to coexistence as I have done through most of this chapter.[73] But this also implies major changes in the ways in which communities interact with authorities. The increasing militarisation of wildlife authorities as a result of the spike in wildlife crime has sometimes had the perverse result that staff are better trained to fight than to negotiate, leading to avoidable escalation of conflict when a problem occurs. Co-existence can be helped by better control methods and technology but also requires a major realignment of priorities and capacity amongst protected area rangers, police and others responsible for managing wildlife.

Dispossession of people from lands, waters and resources

A more intractable issue relates to the role of conservation in forcing people off their traditional territories to set up national parks, wilderness areas, wildlife reserves and other protected areas.

Given the importance of this debate, it is worth outlining exactly what we mean by area-based conservation. There are two main approaches: protected areas and other effective area-based conservation mechanisms (OECMs)

The Convention on Biological Diversity (CBD) defines a protected area as *a geographically defined area which is designated or regulated and managed to achieve specific conservation objectives*.[74] The International Union for Conservation of Nature (IUCN) has a different definition, which the CBD recognises as being equivalent: *A clearly defined geographical space, recognised, dedicated and managed, through legal or other effective means, to achieve the long-term conservation of nature with associated ecosystem services and cultural values*.[75]

These definitions are guidelines: the details of what does and does not "count" as a protected area are determined by national policy and laws. Both IUCN and the CBD recognise a range of management approaches (Table 6.1) and governance types (Table 6.2) as being suitable for protected areas, as long as the areas also meet the definition of a protected area.[76]

In 2010, Aichi Biodiversity Target 11 from the CBD invented a new phrase and started a decade of debate about its implications: "By 2020, at least 17% of terrestrial and inland water areas and 10% of coastal and marine areas ... are conserved through ... systems of protected areas **and other**

Table 6.1 IUCN and the CBD recognise several different management categories

Ia	Strictly protected areas set aside to protect biodiversity and also possibly geological/geomorphological features, where human visitation, use and impacts are strictly controlled.
Ib	Usually, large unmodified or slightly modified protected areas, retaining their natural character and influence, without permanent or significant human habitation
II	Large natural or near natural areas protecting large-scale ecological processes, which provide a foundation for environmentally and culturally compatible spiritual, scientific, educational, recreational and visitor opportunities.
III	Set aside to protect a specific natural monument, which can be a landform, sea mount, submarine cavern, geological feature such as a cave or even a living feature such as an ancient grove.
IV	Aim to protect particular species or habitats and management reflects this priority.
V	An area where the interaction of people and nature over time has produced an area of distinct character with significant ecological, biological, cultural and scenic value: and where safeguarding the integrity of this interaction is vital to protecting and sustaining the area and its nature conservation and other values.
VI	Established to conserve ecosystems and habitats, together with associated cultural values and traditional natural resource management systems.

Table 6.2 Governance types: IUCN and the CBD recognise four governance types of protected areas

A	A government body (such as a Ministry or Park Agency reporting directly to the government) manages the protected area and determines its management aims and objectives.
B	Complex institutional mechanisms and processes are employed to share management authority and responsibility among a plurality of (formally and informally) entitled governmental and non-governmental actors.
C	Protected areas under individual, cooperative, NGO or corporate control and/or ownership set up and managed under not-for-profit or for-profit schemes.
D	Includes two main subsets: (1) indigenous peoples' areas and territories established and run by indigenous peoples and (2) community conserved areas established and run by local communities.

effective area-based conservation measures…" (my emphasis).[77] This led, after considerable work, to the parties to the CBD agreeing a definition of a so-called OECM in Sharm el Sheik, Egypt in late 2018:

A geographically defined area other than a Protected Area, which is governed and managed in ways that achieve positive and sustained long-term outcomes for the in situ conservation of biodiversity, with associated ecosystem functions and services and where applicable, cultural, spiritual, socio–economic, and other locally relevant values.

This covers three main cases:

Primary conservation – areas meeting the IUCN definition of a protected area, but where the governance authority (i.e., community, indigenous peoples' group, religious group, private landowner or company) does not wish the area to be reported as a protected area.

Secondary conservation – active conservation of an area where biodiversity outcomes are only a secondary management objective (e.g., some conservation corridors).

Ancillary conservation – areas delivering in-situ conservation as a by-product of management, even though biodiversity conservation is not an objective (e.g., some military training grounds).[78]

As of December 2022, a third category of area-based conservation was recognised, although it is at the time of writing still too early to see how this will play out in practice. The Global Biodiversity Framework from the CBD noted *recognizing indigenous and traditional territories, where applicable*[79] in its targets for area-based conservation, opening up the possibility (still being discussed) of some Indigenous peoples' territories outside protected areas and OECMs being recognised under CBD targets and therefore effectively being part of "area-based conservation". These changes and potential changes are in part a reaction against conventional protected area models.

Indeed, the vision of protected areas as carefully conserved examples of pristine ecosystems has taken quite a knocking. Critics say that millions of people have been dispossessed by protected areas, stretching back to the First Nations people moved to make way for early national parks like Yellowstone in the United States and the people moved in the establishment of Kaziranga in India. Dispossession ranges from being physically moved from their homes to losing access for the collection of foodstuffs, building material, grazing lands and sacred sites.[80,81] All of these things have certainly happened. We've discussed them in some detail before.[82] Critics argue not only that this is a direct contravention of human rights but that it also doesn't make sense, because the fact that an area has high enough biodiversity value to warrant being a protected area means that the people living there must have been doing the right kind of management anyway – many supposedly "wilderness areas" have been settled for hundreds or thousands of years.[83]

"Officially" this should no longer happen. The CBD has clear guidelines on the need for Free, Prior and Informed Consent (FPIC) before establishment of a protected area and IUCN has agreed principles that protected areas should not be set up through dispossession of people from their lands and waters.[84] Most conservation NGOs now have clear safeguarding policies to avoid human rights abuses.

Forced relocation is used by almost all governments, to build roads, railways, airports, hospitals, industrial and commercial premises. Sometimes this happens on a huge scale, with hundreds of thousands of people involved. The extent to which people are consulted and the level of compensation they receive varies enormously. In theory today the establishment of area-based conservation has less "right" to move human populations under international agreements than a shopping development. Relocation, if it occurs at all, should only be with FPIC – in other words the people moving do so because they want to, with their eyes wide open and with the opportunity to refuse – and they should receive fair and adequate compensation. It is heartening to see more collaborative exercises emerging, where Indigenous peoples and local communities are taking the initiative to conserve[85] or are at least willing partners in carefully negotiated conservation agreements.[86] But it is clear that the rules are not always followed and that bad things still happen in the name of species conservation.

This is almost certainly the most difficult issue relating to human rights and biodiversity rights. When, if ever, is it "right" move people in order to protect other species? The issues are not simple. Quite apart from the human rights perspective, pushing people out of an area that they have been managing sustainably can backfire badly, removing the incentive for good management and creating a surge in poaching, making things worse than they were to start with.[87] Conversely, in other situations protected areas have been set up to allow recovery because it was clear that biodiversity and ecosystem services were declining due to over-exploitation, for whatever reason, this has particularly been particularly studied in marine protected areas.[88] Even when an area has been managed successfully for generations, recent factors such as climate change, human in-migration and out-migration, demographic change and changes in societal expectations can all upset what has previously been a balanced ecosystem. Introduction of guns into an area can radically increase the efficiency of hunting and lead to a net decline of species – the "empty forests" syndrome.[89] We have seen examples of all these situations in our own work.

While it is clear that in many cases humans and other species can co-exist successfully, such as in the conservancies that are being developed in a growing number of African countries, this also sometimes assumes that people will wish to continue a relatively basic lifestyle into the future. There are also well-documented examples of cases where wildlife is declining due to human or livestock pressure in both developed and developing countries. On the other hand, the evidence for successful co-existence is increasing. Wolves, lynxes and bears are all currently expanding in western Europe, three of the trickiest animals to co-exist with and in one of the most heavily populated and developed parts of the world. The success with which humans and other species can co-exist will therefore vary with time and place and simplistic generalisations seldom match reality.

OECMs are supposed to be more flexible and because they are "recognised" from existing management systems should not imply any change in ownership or control; if they work, they work.[90] But given the long tensions about protected areas, many Indigenous peoples and local communities remain deeply sceptical of OECMs as well.

If anyone tells you they are certain that they know the answers to some of the questions above, treat them with extreme caution. We are all still learning our way and old certainties are being challenged at every step. Clearly the top-down, heavy-handed approaches to conservation often used in the past have been shown wanting, both in terms of unacceptable human rights impacts and regarding the long-term security of the ecosystems and species in place. They are increasingly politically impossible in any case, as we'll see when we look at the new international agreements in place towards the end of this book. This won't stop some governments from trying but it will certainly deter NGOs, funding bodies and others from getting involved. New philosophies, like landscape approaches where many stakeholders come together to work out a mutually acceptable landscape and seascape mosaic, have great promise.[91] But they also are new and still not operating at anything but a small scale. And what about protected areas that exist already? Should we be changing management? The issues of restitution are coming to the fore. The US national parks system allowed Native American groups to collect medicinal herbs in protected areas recently, on a sustainable basis. A few protected areas have switched management over to Indigenous peoples or local communities, should this go further? What happens if the new owners want to convert to other uses? While the rich countries have all done their land use changes in the distant past, should we now be demanding that poorer countries preserve their lands intact in order to help mitigate the climate change that we created ourselves? And, coming back to the question being addressed in this book, to what extent should local human communities have the right to change their environment if it threatens the survival of other species?

This last question relates to *who* decides. There is a tendency to assume that people living in an area have the casting vote – or maybe even the only vote – in deciding the management of that area. But where does that leave the rest of the world, including many people who have strong opinions about nature, species and environment? And the many people who don't have access to land and resources themselves but who care about the future of the planet? Whilst I understand that the push for local empowerment is a necessary reaction against the past where local communities were often given no voice at all, giving what will often be small minority complete decision-making power over the future of species or ecosystems also raises major ethical concerns because it assumes that one human group "owns" that species.

Experience suggests that local people often, probably usually, are concerned to maintain a living environment and are supportive of conservation efforts if they can be reasonably integrated with human wellbeing. Most of the most profound changes to our environment in recent decades have come from people based far away, the powerful agribusiness, mining and energy businesses, run by those who will probably never set foot in the places they are exploiting. These stakeholders require different tactics and here the interests of local communities conservationists are likely to agree. But that doesn't mean there will never be clashes between local and global desires regarding biodiversity rights.

Countries where people have been pushed out in the name of conservation are almost always also places where dispossession has occurred for other reasons as well, such as colonialisation, and land-grabbing by large farmers, plantation owners and miners, with the active or passive collusion of governments. Changing these tendencies will be a long-term process. But conservation NGOs and donors need not be complicit in bad practice but rather can use their influence and funding to help seek a more equitable pathway. A mixture of innovation, compromise, participation, proper compensation for any rights foregone, the ability to change minds and assumption of responsibility by the whole global community is needed. In the penultimate chapter some pointers will be suggested. But we are still a long way from having a perfect model to balance human needs with those of the rest of biodiversity.

SHOULD WE EAT MEAT?

We are visiting a major conservation NGO when we are ushered rather peremptorily into a small office where two clearly furious staff members, previously unknown to us, proceed to give us a dressing down the like of which I've not experienced since I was a schoolboy. Our crime? We had co-authored a report that included models of possible futures and their impact on forest cover.[92] The analysis by our collaborators, the highly respected International Institute for Applied Systems Analysis based outside Vienna, concluded that the quickest and most direct route to reducing pressure on tropical forests was for people to eat less meat. Not *no* meat necessarily, even a slight downturn in global consumption would, according to IIASA, have identifiable and positive effects. But our colleagues vehemently disagreed. Meat eating is good for the environment; they're working with the industry to promote more sustainable systems; our analysis is completely off track. We're sorry to have caused offence but know that other people in the same organisation think the opposite. Two completely different world views are clashing here.

Until recently the question of whether or not to eat meat, for those who could afford it, was almost wholly about two issues: the ethics of killing sentient beings for food,[93] and the perceived health benefits of eating or not eating meat.[94] For some, the concerns extended to any animal products. Not so long ago, at least in the west, vegetarians and vegans were a small minority, an identifiable group. Trying to research precise statistics for this section has proved difficult, with widely varying estimates from different sources, with India generally thought to have the highest number of vegetarians at around a quarter of the population but many countries still reporting only 1 or 2 per cent. But what is definitely changing is the number of people who are eating vegetarian or vegan options on an occasional or regular basis, in other words are reducing meat and dairy consumption – sometimes known as "flexitarians". Drivers here are probably more about health and environment than animal welfare as such although flexitarians are also likely to invest more in their occasional meat products and avoid factory farmed or other intensive systems. The US plant-based food market was worth $7.4 billion in 2021,[95] and has recorded regular annual increases for several years running. While this is still a tiny proportion of the total market, it marks a definite and continuing societal shift.

There is a large and fascinating literature on the ethics of eating meat and animal products. Decisions come down to what feels comfortable for each one of us, or what the dictates of a particular faith or philosophy or culture advise. However, this isn't our subject here; this book is not about the rights of individual animals. The question is about whether eating animals is bad for biodiversity *as a whole*, or for the survival rights of individual species.

There is little serious debate that the thriving and largely illegal trade in bushmeat involving rare species is bad for the species involved, with illegal killing being identified as a globally significant threat to many species. Analysis published by the Royal Society in London estimated that 301 terrestrial mammal species were threatened by extinction at least partly due to the bushmeat trade.[96] Given that many of the hunters supplying this trade are amongst the poorest members of society, and probably following a largely traditional lifestyle, the conflict here is between biodiversity rights and human rights, meaning that responses need to be nuanced, working with communities to find sustainable solutions.[97] But while the role of bushmeat hunting as a threat to biodiversity rights is generally accepted, the impacts are limited to a small proportion of consumers, primarily subsistence hunters and a proportion of the aspirant middle classes in developing countries. Education of these consumers is clearly a priority but it is not an argument against eating meat as such.

A far more controversial question relates to whether eating livestock or common and/or sustainably managed game animals is bad for biodiversity. Here things get complicated and opinions increasingly polarised.

Let's take livestock production first. On the debit side there are undoubtedly costs. Meat production is almost certainly the largest cause of land use change, including the clearance of tropical forests, which itself has dramatic negative impacts on biodiversity. Forest has been cleared to create much of the new grazing lands in Latin America for instance.[98] Meat consumption around the world has virtually doubled since the 1960s.[99] Meat has become increasingly controversial because of wider environmental costs it creates. The processes involved in producing meat require about five times more land per unit of nutritional value than a plant-based equivalent.[100] The production of animal products has in consequence dominated agricultural land use change, expansion and intensification for the last half century.[101] If animals are kept indoors or in enclosures, relying on feed grown elsewhere, paradoxically the land required may increase even more. While industrial livestock – factory farms – can be an economically efficient way of producing large quantities of animal products, they are an inefficient way of converting solar energy to nutrient-dense food for humans.

Today, most pigs and poultry are kept indoors and rely solely on protein-rich feed and pharmaceuticals to enhance growth.[102] Around 36 per cent of calories produced by the world's crops are diverted for animal feed, over a third, with only 12 per cent of those feed calories ultimately contributing to the human diet as meat and other animal products. This means that almost a third of the total food value of global crop production is lost by "processing" it through inefficient livestock systems.[103] Even animals kept outdoors all year often require supplementary feeding during cold or dry seasons, adding to the total land required to keep them alive. And it is not just the total amount of plant food but also the types of food. Both soy[104] and oil palm[105] are extensively used in animal feed both have a long history of links to destructive land use change and impacts on wildlife.

When the land used for grazing and feed crops is combined, livestock production accounts for around 70 per cent of agricultural land[106] and is perhaps the single largest driver of biodiversity loss.

Cattle have attracted particular attention with respect to their impacts on biodiversity. Beef is the most costly livestock product in terms of its inefficiency and impacts on land use and pollution, requiring an order of magnitude more resources than many other livestock species, although steps can be taken to reduce some of the impacts.[107] On average, beef requires 28 times more land and 11 times more irrigation water; it produces five times the greenhouse gas emissions and six times the reactive nitrogen impacts (sources of pollution and further biodiversity loss) than livestock such as pigs and poultry.[108] Inefficient beef production also drives land use change, and not only in the developing world. In Queensland, Australia, woodland clearance for cattle pasture averaged 300,000–700,000 ha per year through the 1990s.[109] A ban on further clearance was introduced in 2006 but relaxed in 2013 after pressure from farming groups. Surface runoff has increased 40–100 per cent due to deforestation, endangering the Great Barrier Reef, the largest coral system in the world and currently judged to be at extreme risk of further decline. Even this is not enough; while over 40 per cent of Queensland's cropland is devoted to producing cattle feed, additional imported feed is still required, thus exporting biodiversity impacts elsewhere.[110]

Livestock rearing is also, incidentally, the cause of major reductions in ecosystem services. Livestock production uses a lot of water: average water use for maize, wheat and husked rice is 900, 1,300 and 3,000 m^3 per ton respectively; while that for chicken, pork and beef is 3,900, 4,900 and 15,500 m^3 per ton.[111] Livestock production uses a lot of energy and is also a major cause of climate change, producing an estimated 14.5 per cent of anthropogenic greenhouse gas emissions, with feed production, processing and enteric fermentation (farting, releasing methane), being the main sources of emissions; beef and cow's milk production contribute 41 and 20 per cent of the sector's emissions respectively.[112] There is little dispute that reducing beef consumption, for instance, would have immediate and positive impacts on greenhouse gas emissions.[113] Climate change has direct and serious impacts on biodiversity.

Given this litany of problems the response seems clear cut, to reduce or eliminate meat. That is what I've done, radically cutting down on meat products and now eating meat maybe once a week. I've been exploring vegan alternatives, substituting milk with oat drink and learning a whole new set of cooking skills.

And yet, what about those passionate supporters? Social media is choc-a-bloc with opposite opinions. Some of this smells to me like the actions of paranoid sectors of the livestock industry: sneering attacks on vegans, stories about how people started eating meat again and their lives were transformed, exposes of the environmental impact of the alternatives, etc. But there are more complex issues as well. In the absence of natural grazers, eliminated or reduced to rump populations over much of the planet, livestock grazing maintains natural grasslands and savannah, and livestock farming may be a more benign option for biodiversity than the alternative, which is often intensive monoculture production of crops or trees. Many conservation organisations have invested heavily in schemes that are based around sustainable livestock, working closely with ranchers in what is often an uneasy and fragile partnership, hence the strong reaction to calls for reducing meat. WWF's programme in the Northern Great Plains of North America is a typical example.

These partnerships assume the livestock are on natural grasslands. In fact, many grasslands have been subject to conversion and intensification, including changes in fire frequency and intensity[114] and types and intensity of grazing[115]; introduction of non-native grasses[116]; landscape fragmentation,[117] application of agrochemicals[118]; spread of invasive plant and animal species[119]; and air pollution.[120] So, we might say that biodiversity rights are *often* undermined by livestock but not *always* and that in some situations livestock can be a key element in conservation strategies. This might imply people eating less meat, but also that meat will become more expensive; fine in principle but with clear social implications, increasing the disparities between rich and poor. A transition to a smaller, better managed global livestock population is almost certainly needed and will be aided by increasing sophistication of artificial meat products, but it will not be socially or politically easy to achieve.

WHAT ABOUT TROPHY HUNTING?

We're sitting in a tourist shop in Arusha, Tanzania. We've been there for a couple of hours while our friend Jeff Parrish is searching for a particular mask he wants to buy and we're chatting with the owner, a friendly Tanzanian of Indian descent. Sitting nearby are a bunch of loud, slightly drunk and rather overweight men; they're boisterous and making a lot of noise. "Oh them" the owner says in answer to Sue's question. "They're trophy hunters; they've just spent $70,000 dollars in my shop. I'm giving them a free lunch". A snapshot that sums up all my prejudices; rich, unhealthy-looking people paying huge sums of money to shoot big animals and pose beside them trying and failing to look tough. And in a sharp reversal from a time when hunters sat near the top of the social tree, public dislike of hunters is at all-time high in much of the developed world. The backlash when a dentist from America shot a lion nicknamed Cecil, which had wandered outside a game reserve into a hunting area, included near universal condemnation escalating to death threats.[121] Bans on import of hunting trophies are coming into force in an increasing number of countries. And to be clear, this isn't just about the ethics of killing animals. Many hunters don't like that kind of hunter; many Indigenous and other local peoples who hunt for food accompany their activities with rituals that aim to respect the animals being killed,[122] and they often find the idea of hunting for the joy of killing or posing with a dead animal to be abhorrent. Proposals in Tanzania to expel people from their traditional lands to create a hunting reserve for clients in Saudi Arabia caused a minor firestorm, which continues as I write,[123] it also rebounded on the conservation community when the reserve got conflated with a conservation area.[124] There are persuasive arguments that trophy hunting is morally wrong.[125]

And yet. The situation is far from simple. Hunting reserves cover huge areas of southern Africa and elsewhere, and paradoxically their management is often more effective than in the cash-strapped protected areas that they frequently border.[126] They provide money that helps address the current massive shortfall in conservation funding.[127] Apart from the individuals that are hunted, the rest of the ecosystem is often in good shape and in places where it would almost certainly otherwise have been snapped up for monoculture crops or mining interests. And if countries in Europe and North America are so concerned about hunting, why do they continue to promote it at home? Just looking at my own country, the impacts of maintaining artificially high populations of red deer for hunting and pheasants for shooting are almost wholly negative on the ecosystem as a whole, in marked contrast with game hunting in Africa. Management of artificially high deer populations in Scotland means that most of the country is denuded of woodland. The total biomass of pheasant and partridges released every year in the UK exceeds the total biomass of the wild bird population,[128] creating sharp and increasing imbalances in ecology.[129]

A few brave scientists have stuck their necks out and said that – at the present time – a total ban on trophy hunting in Africa would be catastrophic for wildlife.[130] They've received the depressingly predictable onslaught of abuse, insults, rape threats and death threats from people calling themselves animal lovers. But from the perspective of biodiversity rights – in the sense of survival of species and ecosystems rather than the rights of individuals – hunting is extraordinarily complicated. Some of the highest profile issues, which most people interested in conservation instinctively recoil from, are less of a problem than some of the traditional sports that tend to slip under the radar. For sure, hunters blasting away indiscriminately at migratory raptors in Malta are undermining the survival odds of species[131] and the efforts to halt this pastime is wholly in line with what I'm talking about here. Similarly, the men (and it is usually men) in southern Europe who think it sport to shoot as many songbirds as possible are damaging not only the bird species but also the ecosystems of which they are a part.[132] But managed hunting of chamois in Switzerland, or the killing of wild boar in Germany (where in the absence of predators they can reach damagingly high populations) is either not doing any harm to the ecosystem or potentially helping to maintain some kind of ecosystem balance. Indeed, there are claims that, in the absence of other predators, hunting is the main control of wild boar and declining numbers of hunters is creating damaging levels of wild boar populations in some areas.[133] In Britain, a managed process of roe deer hunting for food might help forests recover in areas where they are over-grazed.

There is a broader question in situations where hunters replace natural predators as to whether this illustrates more fundamental issues relating to ecological functioning. But the overall lesson is that from a biodiversity rights perspective hunting pressure needs to be judged very much on a case-by-case basis. I know passionate conservationists who go out and shoot animals for food, which they argue is a better option than, for example, buying meat from an animal or bird raised on a factory farm. Many hunters deeply respect the animals they hunt and there is a long history of hunters also being supporters of conservation. I don't really understand the mentality of sport hunting although I've talked to many hunters and recognise that at least some of them have deeply felt philosophical and cultural reasons for what they do. Many of the points made here apply equally to sports fishermen, both freshwater and marine, where the same debates are ongoing.

This doesn't mean I like those big game trophy hunters strutting around Nairobi and Dar es Salaam in their expensive safari clothes, who all the locals laugh about. I don't like the people who helicopter into the valley near me in Wales and spend a lot of money to shoot a fat and suicidally tame pheasant. I can't begin to understand how causing its death somehow qualifies as a sport even for – particularly for – someone committed to blood sports. Nor do I see either of those activities as something that we should be encouraging as an essential or acceptable part of human culture. These are valid and important issues for *animal* rights. But the impact of the two examples quoted immediately above on *biodiversity* rights is very different.

Over-stocking of pheasants and partridges is damaging the UK ecosystem whereas hunting managed populations of lions in Africa probably is not. So, while I'm waiting impatiently for the UK conservation organisations to launch their long-promised campaign against pheasant and partridge shooting at its current scale, I remain ambivalent about campaigns against trophy hunting because of the likely consequences on overall survival of species and ecosystems. To be clear, I don't like trophy hunting. But simply eliminating it without alternative land management systems in place will probably result in a much bigger loss of biodiversity.

ARE ALL SPECIES EQUAL?

There's a scene in the movie *Pulp Fiction* where the characters played by John Travolta and Samuel L. Jackson are arguing about food. I paraphrase. Jackson says he doesn't eat pork. Travolta asks if it's because he's Jewish. No, but pigs are filthy animals and eat their own faeces. Travolta points out that dogs eat their own faeces and are not generally considered filthy animals (although definitely dirty) and there follows an expletive-laced discussion about the relative characters of dogs and pigs and whether if a pig was more charming it would somehow cease to be a filthy animal.

We have discussed the illogic of people's attitudes to different animals earlier. Looked at more generally, are all species created equal? From the perspective of biodiversity rights, our definition appears to make it clear that everything is important, there are no favourites. But alongside the cultural, philosophical and spiritual aspects to this question there are also some practical ecological questions to address.

First, what exactly do we mean by "everything"? Earlier discussion has already shown that the definition of species isn't clear cut. If, as argued, biodiversity rights as opposed to animal rights are about the right to continue along certain genetic pathways, this suggests that from an evolutionary perspective, very similar species are possibly less important, because their loss does not cut off many future options, and therefore their "rights" may be less significant than widely divergent species. The loss of one of those ten identical-looking skipper butterflies in Costa Rica will not make much of a dent in the planet's genetic inheritance when compared with the extinction of the Yangtse river dolphin.[134] Philosophically, this kind of reasoning could drop us into a very deep, black hole, but the distinction is not a trivial one.

Then again, from the point of view of the rights of an ecosystem to continue to exist, the answer to whether the species it contains are equal has to be a resounding no. Some species can slip into extinction causing scarcely a ripple, whatever the status of their ethical rights, while the loss of others creates huge changes to ecosystems. Many small species with a limited range, or species whose ecological niche can be quickly filled by others, can disappear and it will be hard to notice they are gone; indeed, these kinds of losses are happening all the time, often of obscure species as yet undescribed by scientists. Conversely, at the other end of the spectrum loss of some species – often referred to as *keystone species* – can have a dramatic trickle-down effect that impacts on whole ecosystems. These are usually either carnivores at the top of the food chain, or herbivores large enough to make substantial ecosystem changes, but can be other, smaller species that have critical roles in shaping the way in which an ecosystem functions. Some will scarcely be known, like particular fungi and microbes that can have dramatic impacts in terms of nutrient cycling or ecosystem perturbation, for instance through the spread of novel diseases.

The fact that over much of the planet the largest carnivores and the most powerful herbivores became extinct a few thousand years ago, probably due to human intervention,[135] means that many of the expected vegetation patterns are almost impossible to achieve without human intervention, and many animals have distorted populations. In evolutionary terms these losses happened yesterday, nature has had no time to adjust. Anyone who has watched elephants casually knocking over trees in the African savannah will have an idea about what loss of the mammoth has had on forest structures in the temperate regions.[136] Loss of bear and wolf over much of North America has created a plague of white-tailed deer, undermining forest regeneration.[137] What we assume is "natural vegetation" in some of the world's wildest places has, in many cases, been radically altered by quite recent changes. The "wilderness" that white settlers discovered when they finally reached the west coast of North America was in large part caused by the massive rate of death of the original inhabitants, killed by imported European diseases that travelled far more quickly than the descendants of the Pilgrim Fathers.

Plants show similar differences. Some long-established species, such as trees, develop multiple hangers on; parasites, mycorrhizal fungi, caterpillars, nesting birds, and later woodpeckers, boring beetles and saprophytic fungi that feed once the tree dies. Other species, confined to a single mountain or island, will have less interactions – but the interactions they do have are more likely to be unique.

A narrow focus on keystone species can miss the wider picture. Obscure, range-limited species are not unimportant from an ecological perspective.

Quite apart from their intrinsic rights to continue to exist as an individual species, they also contribute to the genetic pool that maintains the potential for further evolution, which will help ecosystems adapt to changing conditions. This is particularly important given the extent to which we are altering the ecology of the planet. Apparently obscure species can sometimes play important, hitherto unrecognised roles in an ecosystem functioning.

Notwithstanding the caveat above, from the wider perspective of overall biodiversity rights, the knock-on effects of losing keystone species are likely to be far greater than the impacts of losing some other species and a disproportionate effort on their conservation is justified. In a world of huge threat and limited resources we often have to follow a form of triage, focusing attention on those species that play the most critical roles in an ecosystem. Some of those will be easy to spot; we know that the loss of the huge herbivore migration in the Serengeti-Mara ecosystem of Tanzania and Kenya would have profound impacts on the ecosystem.[138] Others will be much harder to recognise. Researchers initially believed extinction of the dodo on Mauritius would cause loss of the tambalacoque tree (*Calvaria major*) due to digestion in the bird's gut being needed for seed germination.[139] In fact, trees are still being grown successfully,[140] and irreplaceable mutualism disproved.[141] Even in well-studied cases, establishing links is difficult (and romantic stories will tend to be perpetuated whatever the facts).

In practice, the situation is often confused because cultural values get mixed up with conservation values. Species that people care about, either because of the way that they look, or their historical and cultural associations, get a disproportionate amount of conservation attention. Working in what is quintessentially a global field, it is surprising how few conservationists look seriously beyond their borders. Disproportionate amounts of time are spent in the conservation of high-profile species in countries where they are on the edge of their range, and therefore always vulnerable, even if they remain common elsewhere. Daniel Lim and colleagues recently analysed priority species for conservation in the UK. They compiled lists of 187 globally threatened species and 661 endemic species to the UK and found that 55.1 per cent of globally threatened species and 87.5 per cent of the UK's endemic species were not on UK priority species lists. Instead, attention is focused on species already at the edge of their range and common elsewhere, such as the osprey (*Pandion haliaetus*), common crane (*Grus grus*) and the large blue butterfly (*Phengaris arion*).[142] I've spent the odd night guarding the osprey eggs in our local nature reserve and it does seem slightly odd behaviour for one of the commonest raptors on the planet.

But this also illustrates a conundrum for conservation; it only has a hope of surviving if civil society is prepared to offer support. The re-emergence of ospreys and cranes is exciting, whereas action to preserve a fish species endemic to one or two lakes, which it takes an expert to tell from many other species, is less of a crowd pleaser. This shouldn't affect our discussion of rights but it does have an important impact on how those rights are defended in the real world. While all species have equal rights, practical considerations will mean that some of the more influential species get a larger share of conservation attention, and cultural considerations will also, inevitably, have an influence as well.

SPECIES RIGHTS OR ECOSYSTEM RIGHTS?

In 2007, I spent a good few months working on an environmental and social impact assessment of a proposed pulp plantation in Uruguay.[143] As far as we could tell the first detailed written records of the Uruguayan ecosystem came from Charles Darwin, but by then the land had been settled for millennia, first by the original settlers and then by European colonists, and most of the land was down to pasture. Rich pasture, with a mass of flower species and healthy populations of birds and mammals.[144] As in many parts of Latin America, this traditional cultural landscape was rapidly being grubbed up for soy plantations[145] and pulpwood.[146] Alongside the impacts on biodiversity,[147,148] there was considerable pushback from the local gaucho culture, the cowboys of the pampas and campos, seeing their traditional lifestyles disappearing behind a mass of trees and crops. Plantations had become mechanised.[149] A single person operating a harvester could fell, debranch, debark and cut to length an average of 55 eucalyptus stems an hour and rural employment was in danger of collapsing. We heard a lot of problems. But one of the most unlikely from my perspective was resistance from farmers because they said that pumas could use the pulp plantations for cover and increase their raids on livestock.[150]

In the plethora of attempts to define an ethical framework for conservation, *ecosystem rights* has also gained some attention: broadly speaking the right of ecosystems to remain both in existence and functioning in terms of energy and nutrient pathways, food webs and other behavioural, chemical and spatial relationships between the component species. Governments are starting to recognise ecosystem rights: for example, the Ecuadorian constitution (especially Articles 10 and 71–74) recognises the *inalienable rights of ecosystems to exist and flourish*. These concepts overlap much more closely with, and for practical purposes are often virtually identical with, biodiversity rights as described here. The definition of biodiversity, refers to diversity of ecosystems, species and genetic variation within species. The definition of biodiversity rights suggested earlier is specific about the need for a "functioning ecosystem". For practical purposes if we take biodiversity rights seriously, it must include ecosystems, if only because species need healthy ecosystems to survive. And if we are interested in all species this implies a full range of ecosystems, including complex, ancient systems not modified or only minimally modified by humans.

There have been disagreements within the conservation movement about the relationship between species and ecosystems, particularly when identifying conservation priorities, where the choice between taking a species-based or ecosystem-based approach raised high emotions for a while.[151] I never really understood these differences, which seemed to dissolve the deeper you took the analysis. A minority of species can exist effectively outside their ecosystem, or in a highly modified ecosystem and some of these will be amongst the commonest and most recognisable species, precisely because they can exist in human dominated landscapes. But in broad terms, the survival of the majority of species relies on the survival of a functioning ecosystem, so the two are synonymous.

However, there is a question about whether individual species retain significant rights even if ecosystems change or disappear. Some species now only survive in captive breeding populations in zoos or similar; or in "wild" habitats specifically set aside for their use. In some cases, we might regard this as, or hope that this is, a temporary situation until surrounding habitat can be restored or a particular threat removed. But climate change and other forms of degradation may make this state of affairs more or less permanent. Some endemic species face the prospect of their island habitats being overwhelmed by the sea in a few decades; for others the presence of alien invasive species has made their original ecosystem unviable. Should such species be left to go extinct? Or settled elsewhere? The latter has already happened for some of the New Zealand bird species that are being driven to extinction by invasive European mammals: the kakapo (Strigops habroptilus) now only survives on three remote offshore and predator-free islands where they have been deliberately introduced; one of these is within swimming distance of the shore for stoats and requires constant surveillance. In 2022 there were 252 birds, the highest population in 50 years.[152] Governments and conservation organisations seem to assume that such actions are relevant and ethically consistent, at least for a proportion of highly threatened species. Maybe one day science will discover a way of eliminating New Zealand's introduced mammal species without causing any knock-on effects, and the kakapo will be able to spread back over the whole of its original territory. But maybe not: and we are still at the very early stages of working out how to respond to such dramatic changes in ecology.

Also, when we start talking about ecosystem rights, what particular ecosystem are we referring to? The old idea of ecosystems being more or less fixed, at least at the scale of human history, has now been large abandoned; everything changes. Fire, treefall, snow damage, sea-level rise, pest and disease incidence, changing river courses and the natural dispersal of species around the world would be enough to encourage a constant dynamic in many ecosystems even if *Homo sapiens* had never evolved. When these natural agents of change are added to the massive disruptive pressures brought about by air and water pollution, land conversion, soil degradation, engineering of watersheds, the artificial movement of species and above all climate change, the concept of preserving "ecosystems" becomes increasingly slippery and vague. Scientists have begun to recognise, and to reconcile themselves with, the idea that we will be living in a world full of some very different ecosystems in a short while. Dubbed "novel ecosystems",[153] these new assemblages of plants and animals will have to find different ways of interacting, and a proportion of their members will probably be unable to adapt and will disappear. So, ecosystem rights cannot just be about preserving an ecosystem in some (mythical) steady state but must somehow embrace ecosystems constantly in flux and even new ecosystems, filled with new species assemblages. This means, to look back at the debates in the last chapter, that the ethical choices of distinguishing between "good" species and "bad" invasive species become even more complicated. In her book *The Rambunctious Garden*, Emma Marais talked to a lot of scientists who argue we should simply be embracing the change; that today's dangerous invasive species will become a settled member of the ecological community tomorrow, with its own predators and parasites.[154] The question of if and when we should intervene is tackled in the penultimate chapter of this book.

While professional philosophers wrestle with these questions it is, as a conservationist, reassuring to see that a government like Ecuador can simply assert this right on behalf of its ecosystems.

DO WE JUST LET NATURE TAKE OVER? THE REWILDING DEBATE

At the edge of our village in Wales there is a big sign in Welsh and English: *Dweud NA I Ailwylltio — Say NO to Rewilding*, placed there by farmers who see rewilding and carbon markets as the latest threat to their livelihoods. There are also bumper stickers around that say, interestingly, *Conservation Yes — Rewilding No!* suggesting that the issue is not nature as such but the conditions in which it exists. A project to "rewild" an area from the top of Plynlimon, the mountain that is the source of the Rivers Severn and Wye, down to the coast has encountered huge opposition, one of a number of battlefields current around the country.

The concept of rewilding emerged in the last few years particularly in Europe and North America, spurred on by some spectacular examples of species recovery once pressures are reduced, and helped by new advances in understanding of ecology. Predictably, there has also been entrenched opposition, particularly from the farming lobby. This particular flashpoint in mid-Wales, illustrated by the billboard,[155] is probably more to do with some missteps by the organisations involved and deep-seated fears in the farming community about the future after Brexit, without the safety net that the European Common Agricultural Policy has provided for the last 50 years. But other projects in the country have also created very mixed responses.[156] The situation globally hasn't been helped by clumsy signalling from the conservation movement, with mixed messages coming from proponents and instances of maps of "rewilding visions" published, covering existing farms and other private property, long before anyone thought to talk to the people who live there. But it is also a signifier of the increasingly polarised positions developing over this particular issue.

The politics of rewilding do not concern us here. A deeper question is whether rewilding is the best way, or a legitimate way, to protect and recover biodiversity rights. More fundamentally still, given that the proposed definition of biodiversity rights includes the right to exist in a "functioning ecosystem", what does this really mean in practice?

Rewilding in its purest form hinges on the recognition that early humans are likely to have caused the extinction of many of the world's largest predators and herbivores (megafauna) and thus seriously disrupted ecosystems over much of the planet. Over the past 130,000 years (an instant in evolutionary terms) 177 of 294 known megafauna species disappeared, with dramatic losses accelerating around 22,000 years ago, coinciding with the time that humans starting dispersing around the planet.[157,158] According to this theory, in Africa where modern humans first evolved, species had time to adapt to growing hunting pressure from intelligent and cooperative groups and extinctions have generally been less, than average. Then the rate of loss starts to increase in Europe, where humans first migrated, and became almost total in the Americas and Australia, where accomplished human hunters met with megafauna that had no time to evolve defensive strategies. In Australia, the hippopotamus-sized diprotodon and the three-metre tall short-faced kangaroo were just two of a huge group of species that went extinct in the centuries after humans reached the continent; in all 21 genera were lost around the time that humans colonised the continent.[159] In North and South America, four species of elephant, two species of giant armadillo, camels, horses, bison and several species of deer disappeared at about the time humans arrived.[160]

The role of humans in these losses has been controversial for years. The extinction wave also coincided with a period of great climatic instability and some researchers believe that successive ice ages may have played a more substantial role than Homo sapiens or our immediate ancestors. The greatest extent of the ice sheet was also around 22,000 years ago. Three hypotheses compete as an explanation for the dramatic scale of loss: hunting, climate and a mixture of hunting and climate: overkill or overchill. The evidence for hunting being a primary cause is much stronger in the cases of mammals that survived for longer periods, often in remote places that humans for a long time did not penetrate, such as the mammoth that only finally disappeared around 2,500 years ago in places like Wrangel Island north of Siberia or on St Pauls Island south-west of the Alaskan mainland.[161] Without labouring the point, the evidence for hunting being a major, if not the major, cause of loss is increasing, although the question is far from resolved.[162,163]

The controversy goes way beyond the fields of fossil hunters and specialists in ecological history. Indigenous peoples have often been presented – and have presented themselves – as societies living in harmony with nature. The message that their ancestors had caused some of the most dramatic extinctions in the planet's history were unwelcome and writers like Tim Flannery who advanced this hypothesis attracted a lot of personal animosity. This doesn't really make sense; whatever happened took place right at the beginning of human society when it is a stretch to imagine that anyone would have understood issues like carrying capacity and maximum sustainable yields, and many tribal peoples have developed very sustainable ways of living in the millennia since. But the issue is hugely sensitive and the accusations still hurt.

For our purposes here, the final conclusions are less relevant than the undeniable fact that most parts of the world lost most of their largest predators and herbivores in what is ecological terms is an instant.[164] When we stand in a "natural ecosystem" in the Amazon, or the vast temperate rainforests of the Pacific northwest in America, or in the tundra of the far north, it feels completely natural, but we are actually in a very altered ecosystem.

Halfway through the writing of this book, three of us spent a few days in Akagera National Park in Rwanda. Driving along the same tracks, we could see daily signs of elephant activity, including whole trees knocked over. Anyone who spends time in elephant country will come across the same thing. Often referred to as a sign of elephant "damage", in ecological terms the fact that a grumpy old male takes out his frustration on a tree is often just what the ecosystem needs. The reasons why the African savannahs are not completely enclosed by forest (and thus missing all the species that are adapted to grassland and savannah) are a combination of wildfire, the impacts of herbivores and, for the last few thousand years, human activity. The mid-twentieth century idea that temperate countries like the UK were at one time completely covered by wildwood doesn't stand up to the evidence – not least the fact that thousands of plants and animals have evolved over millennia to live in open habitats. Until a few thousand years ago mammoths and in particular the similarly extinct straight-tusked elephants would have been doing much the same as the elephants of Africa, blundering into trees, ripping off branches, creating gaps and generally maintaining the diversity of a mixed habitat. Occasional lightning-strike fires would have cleared larger areas. Some tree species have difficulty reproducing under themselves because caterpillars kills the saplings. The mosaic will have been more variable than previously assumed.

This also means that the early ecological theory of ecosystems heading inexorably towards a "climax vegetation" have been replaced by the concept of a mosaic of different vegetation types, tending towards forest in many places but periodically grazed, battered or burnt so that grassland dominates for a while. The role of herbivores is even more important in damp, temperate climates where wildfire is less common. By removing the megafauna, we have taken away a critical process by which nature regenerates. If the wildwood had ever existed on the scale once thought in Europe, it would have been a human artefact.

So, do we have the wrong baseline? Starting in North America, radical ecologists have begun questioning whether it is possible ever to restore a properly functioning ecosystem in the absence of many of the keystone species that shape the environment. They suggested that *conservation should use ecological history as a guide to actively restore ecological and evolutionary processes*.[165] And by "history", they were not talking about the date the Mayflower first landed European settlers on the continent – until then "ground zero" for most ecologists – but the time before the loss of the megafauna. Their *Pleistocene manifesto* also argued that, in the absence of the actual megafauna, it might be better to introduce something that approximated than attempt to conserve an ecosystem that is inherently out of balance.[166] In particular, they suggested substituting the African cheetah for two extinct North American cheetahs to help control deer and other species. In Britain, George Monbiot has discussed introducing substitute species as wide ranging as the elephant and rhinoceros as a substitute for our lost megafauna, whilst recognising the serious practical and social issues involved in such an exercise.[167]

Maybe the idea is more revolutionary in North America than it would have been in Europe. North American conservation has made great play on wilderness as a concept (albeit a word that is hated by many of the Indigenous people that have lived in and managed such "wild" areas for millennia). European conservationists, working in an environment with a history of more obvious management, have long substituted domestic grazing animals for the aurochs and other wild grazers that have disappeared. But conversely many nature reserves in Europe are heavily managed, often to create the conditions for a single target species or group. While such emergency tactics have reversed the decline of some endangered species, they are labour and money intensive and – by focusing on one or a small number of species – risk distorting the ecology and missing other species altogether.

Dramatic though it undoubtedly was, the Pleistocene rewilding is one of three major disruptions that humans have had on the planet, the last two of which have overlapped. The emergence of agriculture resulted in another dramatic set of changes, creating habitats that initially favoured some species while reducing others, and permanently altering entire ecosystems.[168] Since 1945, industrialisation of agriculture has dramatically reduced the range of species that can co-exist. Lastly, the Industrial Revolution has created a wave of physical and chemical changes to the environment, of which the most extreme to date is climate change. The discovery of DDT residues in the body flesh of penguins in Antarctica marked for many recognition that no inch of the planet was unaffected by human activity and the concept of the Anthropocene was born: the human dominated era of the planet.

Rewilding is a new and therefore still fairly malleable concept with a variable baseline; it has been applied to attempts to roll back all three of the great changes. But almost all serious forms of rewilding involve the reintroduction of herbivores and/or carnivores to replace lost megafauna, either substitute species or domesticated breeds. Restoration of small islands in Mauritius brought in giant tortoises from Aldabra atoll in the Seychelles to replace a similarly large and extinct tortoise that had previously grazed the island vegetation. The Oostvaardersplassen (OVP) experiment in the Netherlands, rewilding an abandoned polder, brought in roe and red deer from Scotland, Heck cattle from Germany (the nearest the researchers could find to the extinct auroch) and horses as substitutes for the tarpan. Britain's most famous rewilding project, the Knepp estate in West Sussex, is a 20-year experiment in rewilding a failing dairy farm and has resulted in a resurgence of wildlife through careful use of smaller numbers of traditional livestock such as Exmoor ponies and old English longhorn cattle and bison.[169]

It is not only the farming lobby that has issues with the concept. Government and many ecologists remain sceptical about reintroductions, while there is also a movement of guerrilla rewilding, with amateurs or enthusiasts clandestinely releasing extirpated species back into their traditional homes. Wild boar and goshawks are now back in the UK due to accidental or unofficial deliberate release and the reintroduction of the beaver has been a mixture of negotiation and unofficial action.

From our perspective here, rewilding is relevant for two rather different reasons. First is the question of whether it is a way to help restore ecological balance and to build populations of species that are otherwise in danger of disappearing, and here the answer seems to be a fairly clear affirmative: nationally rare species like turtle doves and nightingales have returned to Knebb farm without any further encouragement. Many other rewilding projects report similar successes. Restoration, of at least something that approximates to the ecosystem before degradation, often occurs more quickly than the pessimists assume. Species recovery, including recovery of endangered species that have been squeezed out of the modern, managed, countryside are a feature of all successful rewilding projects.

But the second is more complicated. If biodiversity rights means that species live in a functioning ecosystem, what does this really imply in practice? Given how much we have altered ecosystems already, would walking away from management simply result in the dominance of a handful of highly successful species with others driven into extinction? Or would things settle down again with most species in their "proper places"? Everyone recognises that unmanaged ecosystems are likely to be more unstable, indeed that is partly the point. But when species are already at the edge of extinction, there isn't much wiggle room to help them survive bad periods. Climate change is already proving to be a huge and complicated agent of change. I have a more-than-sneaking fear that the gung-ho introduction of substitute species will often make things worse rather than better. There is a long and dismal history of human meddling in ecosystems, ranging from the acclimatisation societies that sought to establish European wildlife species in colonised lands[170] to a range of failed biological controls, most notoriously in the case of the cane toad in Australia.

Rewilding is therefore in principle fully aligned with biodiversity rights, but only if it is done carefully. Like biological controls, and other forms of human meddling in ecosystems, rewilding could be either a positive or negative force for change. I have spent a lot of time worrying these issues through: how do we define natural in a world that has been so heavily modified by humans? A decade ago, I proposed the concept of authenticity as a twenty-first century alternative to *natural* by which I mean roughly a diverse and relatively stable system that functions over time. The definition suggested was: *a resilient ecosystem with the level of biodiversity and range of ecological interactions that would be predicted as a result of the combination of historic, geographic and climatic conditions in a particular location.*[171] I would still hold to this, and the inclusion of "historical" means that it could apply in principle to the most radical rewilding concepts. But the way that we do rewilding, and particularly the safeguards built into any experiment, probably need more thought than they have had until now.

WHAT ABOUT NON-NATIVE SPECIES?

Three of us are hanging around a churchyard in Wales, trying to find a family of hawfinches that I know lives there. A man walks down the road carrying a grey squirrel in a cage. The squirrel is very agitated; it knows it is in trouble. Our new friend tells us it is the 27th he has trapped and he is off to drown it in the local ditch. He's a wildlife lover and fills us in about all the birds in the neighbourhood; he explains that he is drowning the squirrel because the native red squirrel has started to make a tentative reappearance in the area after virtually disappearing for decades, and he's busy clearing out the opposition.

But was he right? Grey squirrels, a North American native, were released in the 1930s and 1940s, and despite sporadic attempts at control spread out rapidly through much of England and Wales, reaching the latter in the early 1960s. Red squirrels virtually disappeared soon afterwards.[172] Maybe they had one of their periodic population crashes[173] and the greys filled the gap, perhaps there was active competition; experts disagree.[174] Many people love the grey squirrels; they've never known any other kind and you'll see people in virtually any area of parkland busy feeding them nuts or scraps of bread. On the other hand, where red squirrels still exist, in parts of Scotland, in the Lake District and Northumberland and on isolated areas like the Isle of Wight, they are held in high regard by many local people and are a draw for tourists.[175] The significance is largely cultural; red squirrels are in no danger of extinction on a global basis and still have healthy populations in parts of Britain. The reintroduction and expansion of the pine marten is likely to aid the red; greys are generally heavier than the reds and in consequence are killed preferentially by martens because they cannot escape to the smaller branches that will not support the weight of the predator.[176] But should we be interfering at all? In Italy protests by animal rights activists delayed eradication programmes for a small group of feral grey squirrels so that the grey is now "out of control" there as well[177] and the native red is declining. And if the decision is made to reduce the population, is it better to reintroduce native pine martens to control greys or trap them ourselves?

When I was a boy, brown hares were assumed to be a natural part of the UK fauna, having loped their way from mainland Europe before the English channel flooded with the melt following the last Ice Age. They were certainly present in previous inter-glacial periods. Now archaeologists are fairly sure they were introduced this time around.[178] Does that make me look differently at the hares that leap out from underneath my feet in the hills around my house? It does not.

Ecologists distinguish between non-native species that have settled down peacefully into an ecosystem and those that cause major disturbances, the latter being known as "alien invasive species". There are also so-called "invasive species" that are native, if cultural or other conditions suddenly allow them to outcompete other native species. The bracken fern (*Pteridium aquilinum*) has a cosmopolitan distribution in both temperate and tropical climates. Pasture clearance and excessive burning allows it to spread quickly and outcompete other native plants; it is regarded as "invasive" in many areas of its natural distribution as well as in places where it has been introduced. Some people don't like this term, and believe it has unpleasant echoes of human xenophobia, but it is generally understood and will be used here.

Invasive species are often listed as the second most significant threat to biodiversity. I'd question that, but there are undoubtedly many places, particularly long-isolated islands, where they have had and are having a devastating effect on native species. NGOs like Flora and Fauna International and re:wild run major campaigns to eradicate rats, goats and other species from islands where they are driving endemic species extinct. New Zealand has announced ambitious plans to eradicate stoats, weasels and other introduced mammalian predators that have already driven many species to extinction and confine others to closely guarded offshore islands.[179] Huge sums of money are spent trying to control invasive plants like water hyacinth, that is now blocking waterways almost throughout the tropics.

I have already discussed that eradication campaigns against invasive species can generate a kickback from animal rights activists, even if the non-native species are killing a disproportionate number of other animals or undermining an ecosystem. Intense campaigns have been waged against attempts to control wild horses ("brumbies") in Kosciuszko National Park in Australia, hedgehogs in offshore Scottish islands, American mink in Britain and many more. The intensity of the militant animal rights movement differs enormously, especially strong in parts of Europe and North America it is virtually absent in many other parts of the world. It is clear that this is an area with a great potential for animal rights and biodiversity rights to come into conflict and has been touched on earlier in discussions of the compassionate conservation ethic.

The situation is further complicated because of a lack of unanimity within the field of biodiversity conservation, with a small but significant reaction against the received wisdom that invasive species are generally "bad". As we've discussed, the more hard-line rewilding proponents would be happy to see completely non-native species introduced in places to substitute for others that should be playing a critical role in ecology but are now extinct. More generally, others look at non-native species as one more disrupting factor among many, probably impossible to prevent or eradicate and capable of being absorbed within ecosystems, changing but not destroying them. They argue that outside of unusually fragile ecosystems, especially those on long-isolated islands, invasive species seldom cause extinctions and even if they do there will soon be replacement species within the ecosystem. Some go further and postulate that alien species should be welcomed as the change agents needed in the disruptive conditions created by climate change, so that any losses of endemic species from non-natives will be compensated for by rapid evolution driven by the new genetic mixture.[180] These are far from trivial issues.

To be clear, most introductions are benign to irrelevant, quickly succumbing as a result of unsuitable conditions, or hanging on as tiny populations, or simply being absorbed into the local ecosystem so that they are hardly any more recognised as invasives at all. Like dingoes in Australia, hares in Britain, or Indian myna birds in numerous countries around the world. However, along with the many introduced plant and animal species that simply melt away into their new ecosystem, a minority cause havoc to native species, outcompeting them and sometimes driving them to extinction.

Several factors combine to make a non-native disruptive to local ecology. The best known is that it may have arrived in an ecosystem that has not evolved to control it. Cane toads in Australia are highly poisonous and killed huge numbers of predatory animals in consequence; whole populations of crocodiles were wiped out and it looked for a while as if formerly healthy ecosystems like Kakadu would be virtually destroyed. Nature adapts fast however and survivors learned to avoid the toads, while the smarter species worked out ways to eat them without touching the dangerous poison glands.[181,182] The impact of a new species will therefore be down to how quickly the rest of the ecosystem can adapt. Several of New Zealand's flightless bird species did not have the evolutionary time needed to adapt, meaning that invasive stoats, weasels and foxes wiped them out.[183] The most serious issues arise when the evolutionary changes needed are too significant to occur within the short time necessary (and behavioural changes alone won't do) or when a whole lifestyle is undermined. So, introduction of aggressive species like rats and snakes on islands where they did not exist before poses extreme risks and is why NGOs spend time and money on their eradication.

Another major factor in allowing a native or non-native species to become "invasive" is if they live in an ecosystem that has already been degraded. Bracken spreads if landowners overuse fire to "control" woodland and maintain open pasture.[184] Water hyacinth grows fastest in polluted, nutrient rich water.[185] The invasive jellyfish which clogged the Black Sea arrived at a time when the ecosystem was already severely degraded.[186] In these cases, "control" of an invasive species may be more about improving the overall health of the ecosystem than in targeting the species itself.

So how does this relate to biodiversity rights? The key question is whether non-native species have more or less rights than native species. Use of words like "alien" and "invasive" places this as an issue with wider social and cultural connotations; demonising the outsider is not something to be done lightly in the tense social conditions of the twenty-first century, where issues of human migration split communities and examples of intolerance occur every day.

From the perspective of rights taken here — *to continue their natural span of existence* — then the only time when controlling (which generally means killing or at least stopping the reproduction of) non-native species or invasive native species is justifiable is if they pose a credible threat to the existence of other whole species, genetically significant populations or functioning ecosystems. So brown rats threatening globally significant populations of nesting seabirds on a remote island are fair game. So too, although more controversially, are introduced rabbits killing an endemic plant species on another island. But mass attempts to remove species that have integrated comfortably into ecosystems has no justification from a biodiversity *rights* perspective; it may in some cases have major cultural incentives which can be extremely important, but this is a different issue. Ecosystems are more flexible than was once assumed and can usually absorb a lot of change before undergoing major disruption or species loss even if some of these changes can look quite dramatic.

NOTES

1 Baird Callicott, J. 1989. *Defense of the Land Ethic: Essays in Environmental Philosophy.* State University of New York, Albany: 15–38.

2 https://jbcallicott.weebly.com/introductory-palinode.html

3 Watanabe, S. 2010. Pigeons can discriminate "good" and "bad" paintings by children. *Animal Cognizance* **13** (1): 75–85.

4 Aquinas, St. Thomas, 1922. *The Summa Theologica of St Thomas Aquinas.* Burn, Oates and Washbourne, London.

5 Descartes, R. 1637. *A Discourse on Method.* Oxford University Press, Oxford.

6 Thomas, K. 1983. *Man and the Natural World: Changing attitudes in England 1500–1800.* Allen Lane, London gives a compelling account of changing attitudes in one country.

7 Bergman, G. 2002. The history of the human female inferiority ideas in evolutionary biology. *Rivista bi biologia* **95** (3): 379–412.

8 Sahlins, P. 2017. *1668: The Year of the Animal in France.* Zone Books, New York.

9 Clarke, P.A.B. and Linzey, A. (eds.) 1990. *Political Theory and Animal Rights.* Pluto Press, London.

10 Birke, L. 1986. *Women, Feminism and Biology: The Feminist Challenge.* Harvester, Brighton.

11 Midgley, M. 1983. *Animals and Why They Matter. A Journey around the Species Barrier.* University of Georgia Press.

12 Singer, P. 1975. *Animal Liberation: A New Ethics for Our Treatment of Animals.* Harper Collins, London.

13 Hutchins, M. and Wemmer, C. 1986. Wildlife conservation and animal rights: Are they compatible? In: Fox, M.W. and Mickley, L.D. (eds.) *Advances in Animal Welfare Science 1986/87.* The Humane Society of the United States, Washington, DC: 111–137.

14 Rose, J.D., Arlinghaus, R., Cooke, S.J., Diggles, B.K., Sawynok, W., et al. 2012. Can fish really feel pain? *Fish and Fisheries*. DOI:10.1111/faf.12010

15 Barreto, G.R., Rushton, S.P., Strachan, R. and Macdonald, D.W. 2001. The role of habitat and mink predation in determining the status and distribution of water voles in England. *Animal Conservation* **1** (2): 129–137.

16 See for instance: Singer, P. 1975. *Animal Liberation*. Pimlico, London. Singer argues from a utilitarian viewpoint, which is not shared by all animal rights philosophers, but is probably the book that more than any other launched the "modern" animal rights movement. See also Midgley, M. 1983. *Animals and Why They Matter*. University of Georgia Press, Athens.

17 Muth, R.M. and Jamison, W.V. 2000. On the destiny of deer camps and duck blinds: The rise of the animal rights movement and the future of wildlife conservation. *Wildlife Society Bulletin* **28** (4): 841–851.

18 Regan, T. 1983. *The Case for Animal Rights*. University of California Press, Oakland.

19 For a deeply knowledgeable and rather quirky assessment of cattle see Young, R. 2003. *The Secret Life of Cows*. Faber and Faber, London.

20 Leipämaa-Leskinen, H., Syrjälä, H. and Jaskari, M.-M., 2018. A semiotic analysis on cultural meanings of eating horsemeat. *Qualitative Market Research* **21** (3): 337–352.

21 Dudley, N. and Gordon Clarke, J. 1983. *Thin Ice*. Marine Action Centre, Cambridge.

22 Roberts, K. 1983. *Man and the Natural World: Changing Attitudes in England 1500-1800*. Allen Lane, Harmondsworth.

23 Hamilton-Paterson, J. 2023. *Stuck Monkey: The Deadly Planetary Cost of Things We Love*. Head of Zeus, London.

24 van Aarde, R., Whyte, I. and Pimm, S. 1999. Culling and the dynamics of the Kruger National Park African elephant population. *Animal Conservation* **2**: 287–294.

25 Rolston, H. III. 1995. Duties to endangered species. In: Elliot, R. (ed.) *Environmental Ethics*. Oxford University Press, Oxford: 60–75.

26 US Fish and Wildlife Service. 2007. *Santa Barbara Island Liveforever (Dudleya traskiae): 5-year Review Summary and Evaluation*. Ventura Fish and Wildlife Office, Ventura, CA.

27 Bekoff, M. (ed.) 2013. *Ignoring Nature No More: The Case for Compassionate Conservation*. University of Chicago Press, Chicago, IL and London.

28 Hayward, M., Callen, A., Allen, B.L., Ballard, G., Broekhuis, F. et al. 2019. Deconstructing compassionate conservation. *Conservation Biology* **33** (4): 760–768.

29 Coghlan, S. and Cardilini, A.P.A. 2021. A critical review of the compassionate conservation debate. *Conservation Biology* **36** (1): e13760. DOI:10.1111/cobi.13760.

30 Conservation Assured. 2018. *Safe Havens for Wild Tigers: A Rapid Assessment of Management Effectiveness against the Conservation Assured Tiger Standards*. Conservation Assured, Singapore.

31 Wikramanayake, E., Dinerstein, E., Seidensticker, J., Lumpkin, S., Pandav, B., et al. 2011. A landscape-based conservation strategy to double the wild tiger population: Landscape-based strategy for tiger recovery. *Conservation Letters* **4**: 219–227.

32 Rastogi, A., Hickey, G.M., Badola, R. and Ainul Hussain, S. 2012. Saving the superstar: A review of the social factors affecting tiger conservation in India. *Journal of Environmental Management* **113**: 328–340.

33 Manoj. E.J. 13th September 2021. 40 killed in tiger attacks in the country last year. *The Hindu.*

34 Acharya, K.P., Paudel, P.K., Neupane, P.R. and Köhl, M. 2016. Human-wildlife conflicts in Nepal: Patterns of human fatalities and injuries caused by large mammals. *PLoS One* **11** (9): e0161717. DOI:10.1371/journal.pone.0161717.

35 DoFPS. 2015. *Counting the Tigers in Bhutan: Report on the National Tiger Survey of Bhutan 2014–2015.* Department of Forests and Park Services, Ministry of Agriculture and Forests, Thimphu.

36 Miller, J.R.B., Jhala, Y.V. and Jena, J. 2016. Livestock losses and hotspots from tigers and leopards in Kanha Tiger Reserve, Central India. *Regional Environmental Change* 16 (S1). DOI:10.1007/s10113-015-0871-5

37 Bargali, H.S. and Ahmed, T. 2018. Patterns of livestock depredation by tiger (*Panthera tigris*) and leopard (*Panthera pardus*) in and around Corbett Tiger Reserve, Uttarakhand, India. *PLoS One* **13** (5): e0195612. DOI:10.1371/journal.pone.0195612.

38 WWF Russia. n.d. Tiger reintroduction programme in Kazakhstan. https://wwf.ru/en/regions/central-asia/vosstanovlenie-turanskogo-tigra/

39 Launay, F., Cox, N., Baltzer, M., Tepe, T., Seidensticker, J. et al. 2013. Preliminary study of the feasibility of a tiger restoration programme in Cambodia's Eastern Plains. A Report Commissioned by WWF.

40 Korea National Park Service and IUCN. 2009. *Korea's Protected Areas: Evaluating the Effectiveness of South Korea's Protected Area System.* Seoul and Bangkok.

41 Conservation Assured. 2018. *CA | TS Manual Version 2.* Conservation Assured, Singapore.

42 Hammond, C.E. 1992. Sacred metamorphosis: The weretiger and the shaman. *Acta Orientalia Academiae Scientiarum Hungaricae* **46** (2/3): 235–255.

43 Green, S. 2006. *Tiger.* Reaktion Books, London.

44 Cooper, J.C. 1992. *Symbolic and Mythological Animals.* Aquarian Press, London: 226–227.

45 Reddy, C.S. and Yosef, R. 2016. Living on the edge: Attitudes of rural communities toward Bengal Tigers (*Panthera tigris*) in Central India, *Anthrozoös* **29** (2): 311–322.

46 Krishna, N. 2010. *Sacred Animals of India.* Penguin Books, New Delhi.

47 Verma, M., Negandhi, D., Khanna, C., Edgaonkar, A., David, A. et al. 2015. *Economic Valuation of Tiger Reserves in India: A Value+ Approach.* Indian Institute of Forest Management, Bhopal.

48 Matthiessen, P. and Hornocker, M. 2001. *Tigers in the Snow.* North Point Press, New York.

49 Jackson, P. 1999. The tiger in human consciousness and its significance in crafting solutions for tiger conservation. In: Seidensticker, J., Christie, S. and Jackson, P. (eds.) *Riding the Tiger: Tiger Conservation in Human-Dominated Landscapes.* Cambridge University Press, Cambridge: 50–54.

50 Li, J., Wang, D., Yin, H., Zhaxi, D., Jiagong, Z., et al. 2013. Role of Tibetan Buddhist Monasteries in snow leopard conservation. *Conservation Biology* **28** (1): 87–94.

51 DoFPS. 2015. *Counting the Tigers in Bhutan: Report on the National Tiger Survey of Bhutan 2014–2015.* Department of Forests and Park Services, Ministry of Agriculture and Forests, Thimphu.

52 Ministry of Agriculture and Forests. 2016. *Bhutan State of Parks 2016*. Department of Forest and Park Services, Ministry of Agriculture and Forests, Royal Government of Bhutan, Thimphu.

53 Inskip, C., Carter, N., Riley, S., Roberts, T. and MacMillan, D. 2016. Toward human-carnivore coexistence: Understanding tolerance for tigers in Bangladesh. *PLoS One* **11**: e0145913. DOI:10.1371/journal.pone.0145913.

54 McKay, J.E., St. John, F.A.V., Harihar, A., Martyr, D., Leader-Williams, N., et al. 2018. Tolerating tigers: Gaining local and spiritual perspectives on human-tiger interactions in Sumatra through rural community interviews. *PLoS One* **13**: e0201447. DOI:10.1371/journal.pone.0201447.

55 Mangunjaya, F.M., Elkin, C., Praharawati, G., Tobing, I.S.L. and Tjamin, Y.R. 2018. Protecting tigers with a Fatwa: Lesson learn faith base approach for conservation. *Asian Journal of Conservation Biology* **7** (1): 78–81.

56 Qin, Y. and Nyhus, P.J. 2017. Assessing factors influencing a possible South China tiger reintroduction: A survey of international conservation professionals. *Environmental Conservation*. DOI:10.1017/S0376892917000182

57 Bhatia, S., Redpath, S.M., Suryawanshi, K. and Mishra, C. 2017. The relationship between religion and attitudes toward large carnivores in Northern India? *Human Dimensions of Wildlife* **22**: 30–42.

58 Carter, N.H. and Linnell, J.D.C. 2016. Co-adaptation is key to coexisting with large carnivores. *Trends in Ecology and Evolution* **31**: 575–578.

59 Maginnis, S., Jackson, W. and Dudley, N. 2004. Conservation landscapes. Whose landscapes? Whose trade-offs? In: McShane, T.O. and Wells, M.P (eds.) *Getting Biodiversity Projects to Work*. Columbia University Press, New York: 321–339.

60 Gulati, S., Karanth, K.K., Le, N.A. and Noack, F. 2021. Human casualties are the dominant cost of human-wildlife conflict in India. *Proceedings of the National Academy of Sciences of the United States of America* **118**: e1921338118. DOI:10.1073/pnas.1921338118.

61 See for example: Wang, S.W. and Macdonald, D.W. 2006. Livestock predation by carnivores in Jigme Singye Wangchuck National Park, Bhutan. *Biological Conservation* **129**: 558–565; Miller, J.R.B., Jhala, Y.V., Jena, J. and Schmitz, O.J. 2015. Landscape-scale accessibility of livestock to tigers: implications of spatial grain for modelling predation risk to mitigate human-carnivore conflict. *Ecology and Evolution* **5** (6): 1354–1367; Miller, J.R.B., Jhala, Y.V. and Jena, J. 2015. Livestock losses and hotspots of attack from tiger and leopards in Kanha Tiger Reserve, Central India. *Regional Environmental Change*. DOI:10.1007/s10113-015-0871-5

62 Dickman, A., Macdonald, E.A. and Macdonald, D.W. 2011. A review of financial instruments to pay for predator conservation and encourage human-carnivore coexistence. *Proceedings of the National Academy of Sciences* **108** (34): 13937–13944.

63 Belecky, M., Stolton, S., Dudley, N., Dahal, S., Fei Li, M and Hebert, C. 2022. *Living with Tigers: How to Manage Coexistence for the Benefit of Tigers and People*. WWF International, Gland.

64 Elliott, W., Kube, R. and Montanye, D. 2008. *Common Ground: Solutions for Reducing the Human, Economic and Conservation Costs of Human Wildlife Conflict*. WWF International, Gland.

65 IUCN. 2020. *IUCN SSC Position Statement on the Management of Human-Wildlife Conflict.* IUCN Species Survival Commission (SSC) Human-Wildlife Conflict Task Force.

66 Hodgson, I.D., Redpath, S.M., Sandström, C. and Biggs, D. 2020. *The State of Knowledge and Practice on Human-Wildlife Conflicts.* Luc Hoffmann Institute, Gland.

67 Weise, F.J., Hauptmeier, H., Stratford, K.J., Hayward, M.W., Aal, K., et al. 2019. Lions at the gates: Trans-disciplinary design of an early warning system to improve human-lion coexistence. *Frontiers in Ecology and Evolution* **6**. DOI: 10.3389/fevo.2018.00242

68 Ahmed, A. 2011. *Lighting Lives in the Sundarbans.* WWF-India. wwf.panda.org/?uNewsID=201051 (accessed 6/1/2022).

69 Cavalcanti, S.M.C., Crawshaw, P.G. and Tortato, F.R. 2012. Use of electric fencing and associated measures as deterrents to Jaguar predation on cattle in the pantanal of Brazil. In: Somers, M.J. and Hayward, M. (eds.) *Fencing for Conservation.* Springer, New York: 295–309.

70 Lichtenfeld, L.L., Trout, C. and Kisimir, E.L. 2015. Evidence-based conservation: Predator-proof bomas protect livestock and lions. *Biodiversity and Conservation* **24**: 483–491.

71 Bhattarai, B.R., Wright, W., Morgan, D., Cook, S. and Baral, H.S. 2019. Managing human-tiger conflict: Lessons from Bardia and Chitwan National Parks, Nepal. *European Journal of Wildlife Research* **65**: 34.

72 Goodrich, J.M. 2010. Human–tiger conflict: A review and call for comprehensive plans. *Integrative Zoology* **5**: 300–312.

73 Fisher, M. 2016. Whose conflict is it anyway? Mobilising research to save lives. *Oryx* **50** (3): 377–378.

74 https://www.cbd.int/protected/pacbd/

75 Dudley, N. (ed.) 2008. *Guidelines for Applying Protected Area Management Categories.* IUCN, Gland.

76 Ibid.

77 https://www.cbd.int/sp/targets/

78 IUCN-WCPA Task Force on OECMs. 2019. *Recognising and Reporting Other Effective Area-Based Conservation Measures.* IUCN, Gland.

79 https://www.cbd.int/article/cop15-cbd-press-release-final-19dec2022

80 Colchester, M. 2003. *Salvaging Nature: Indigenous Peoples, Protected Areas and Biodiversity Conservation,* World Rainforest Movement and Forest Peoples Programme, Montevideo and Moreton-in-Marsh.

81 Dowie, M. 2009. *Conservation Refugees.* The MIT Press, Cambridge.

82 Dudley, N. and Stolton, S. 2020. *Leaving Space for Nature: The Critical Role of Area-Based Conservation.* Routledge, Oxford.

83 Fernández-Llamazares, Á., Terraube, J., Gavin, M.C., Pyhälä, A., Siani, S.M.O. et al. 2020. Reframing the wilderness concept can bolster collaborative conservation. *Trends in Ecology and Evolution* **35** (9): 750–753.

84 Dudley, N. (ed.) 2008. *Guidelines for Applying Protected Area Management Categories.* IUCN, Gland.

85 Forest Peoples Programme et al. 2020. *Local Biodiversity Outlooks* 2. Forest Peoples Programme, Moreton-on-the-Marsh.

86 Walker Painemilla, K., Rylands, A.B., Woofter, A. and Hughes, C. (eds.) 2010. *Indigenous Peoples and Conservation: From Rights to Resource Management*. Conservation International, Washington, DC.

87 Ntuli, H., Sundström, A., Sjöstedt, M., Muchapondwa, E., Jagers, S.C. and Linell, A. 2021. Understanding the drivers of subsistence poaching in the Great Limpopo transfrontier conservation area: What matters for community wildlife conservation? *Ecology and Society* **26** (1): 18.

88 See for example Buelow, C.A., Connolly, R.M., Turschwell, M.P., Adame, M.F., Ahmadia, G.N. et al. 2022. Ambitious global targets for mangrove and seagrass recovery. *Current Biology* **32**: 1–9.

89 Bogoni, J.A., Percequillo, A.R., Ferraz, K.M.P.M.B. and Peres, C.A. 2022. The empty forest three decades later: lessons and prospects. *BioTropica*. DOI:10.1111/btp.13188

90 Laffoley, D., Dudley, N., Jonas, H., MacKinnon, D., MacKinnon, K., et al. 2017. An introduction to "other effective area-based conservation measures" under Aichi Target 11 of the Convention on Biological Diversity: Origin, interpretation and emerging marine issues. *Aquatic Conservation: Freshwater and Marine Ecosystems* **27** (S1): 130–137.

91 Dudley, N., Baker, C., Chatterton, P., Ferwerda, W.H., Gutierrez, V., Madgwick, J. 2021. *The 4 Returns Framework for Landscape Restoration. UN Decade on Ecosystem Restoration Report*. Commonland, Wetlands International Landscape Finance Lab and IUCN Commission on Ecosystem Management.

92 Taylor, R. (ed.) 2011. *WWF Living Forests Report*. Chapter 1: Forests for a Living Planet. WWF, Gland.

93 Rachels, J. 2016. The basic argument for vegetarianism. In: Armstrong, S.J. and Botzler, R.J. (eds.) 2016. *The Animal Ethics Reader*, 3rd edition. Routledge, New York: 274–280.

94 Amato, P.R. and Partridge, S.A. 2013. *The New Vegetarians: Promoting Health and Protecting Life*. Springer, New York.

95 https://www.plantbasedfoods.org/2021-u-s-retail-sales-data-for-the-plant-based-foods-industry/

96 Ripple, W.J., Abernethy, K., Betts, M.G., Chapron, G., Dirzo, R. et al. 2016. Bushmeat hunting and extinction risk to the world's mammals. *Proceedings of the Royal Society B* **3** (10). DOI:10.1098/rsos.160498

97 Coad, L., Fa, J.E., Abernethy, K., van Vliet, N., Santamaria, C., Wilkie, D., et al. 2019. *Towards a Sustainable, Participatory and Inclusive Wild Meat Sector*. Center for International Forestry Research, Bogor, IN.

98 Geist, H.J. and Lambin, E.F. 2002. Proximate causes and underlying driving forces of tropical deforestation. *BioScience* **52**: 143–150.

99 FAO. 2013. 'FAOSTAT'. http://faostat3.fao.org/faostat-gateway/go/to/home/E (accessed 11/11/2016).

100 UNEP. 2009. *Towards Sustainable Production and Use of Resources: Assessing Biofuels.* United Nations Environment Programme, Division of Technology Industry and Economics, Paris.

101 Alexander, P., Rounsevell, M.D.A., Dislich, C., Dodson, J.R., Engström, K., et al. 2015. Drivers for global agricultural land use change: The nexus of diet, population, yield and bioenergy. *Global Environmental Change* **35**: 138–147.

102 Schneider, M. 2011. *Feeding China's Pigs: Implications for the Environment, China's Smallholder Farmers and Food Security.* Institute for Agriculture and Trade Policy, Washington DC.

103 Cassidy, E.S., West, P.C., Gerber, J.S. and Foley, J.A. 2013. Redefining agricultural yields: From tonnes to people nourished per hectare. *Environmental Research Letters* **8**. DOI:10.1088/1748-9326/8/3/034015

104 WWF. 2014. *The Growth of Soy: Impacts and Solutions.* WWF International, Gland.

105 Edwards, D.P., Hodgson, J.A., Hamer, K.C., Mitchell, S.L., Ahmad, A.H., et al. 2010. Wildlife-friendly oil palm plantation fail to protect biodiversity effectively. *Conservation Letters* **3**: 236–242.

106 FAO. 2006. *Livestock's Long Shadow: Environmental Issues and Options.* FAO, Rome.

107 Castonguay, A.C., Polasky, S., Holden, M.H., Herrero, M., Mason-D'Croz, D. et al. 2023. Navigating sustainability trade-offs in global beef production. *Nature Sustainability.* DOI:10.1038/s41893-022-01017-0

108 Eshel, G., Shepon, A., Makov, T. and Milo, R. 2014. Land, irrigation water, greenhouse gas, and reactive nitrogen burdens of meat, eggs, and dairy production in the United States. *Proceedings of the National Academy of Sciences* **111** (33): 11996–12001.

109 McAlpine, C.A., Etter, A., Fearnside, P.M., Seabrook, L. and Laurance, W.F. 2009. Increasing world consumption of beef as a driver of regional and global change: A call for policy action based on evidence from Queensland (Australia), Colombia and Brazil. *Global Environmental Change* **19**: 21–33.

110 Siriwardena, L., Finlayson, B.L. and McMahon, T.A. 2006. The impact of land use change on catchment hydrology in large catchments: The Comet River, Central Queensland, Australia. *Journal of Hydrology* **326** (1): 199–214.

111 Hoekstra, A.Y. and Chapagain, A.K. 2007. Water footprint of nations: Water use by people as a function of their consumption pattern. *Waters Resources Management* **21**: 35–48.

112 Gerber, P.J., Steinfeld, H., Henderson, B., Mottet, A., Opio, C., et al. 2013. *Tackling Climate Change through Livestock – A Global Assessment of Emissions and Mitigation Opportunities.* Food and Agriculture Organization of the United Nations, Rome.

113 Stehfest E., Bouwman, L., van Vuuren, D.P., den Elzen, M.G.J., Eikhout, B., et al. 2009. Climate benefits of changing diet. *Climate Change* **95**: 83–102.

114 Pyne, S. 1994. Maintaining focus: An introduction to anthropogenic fire. *Chemosphere* **29** (5): 889–911.

115 Milchunas, D.G., Sala, O.E. and Lauenroth, W.K. 1988. A generalized model of the effects of grazing by large herbivores on grassland community structure. *The American Naturalist* **132** (1): 87–106.

116 Musil, C.F., Milton, S.J. and Davis, G.W. 2005. The threat of alien invasive grasses to lowland Cape floral diversity: An empirical appraisal of the effectiveness of practical control strategies. *South African Journal of Science* **101**: 337–344.

117 Osborne, C.P., Charles-Dominique, T., Stevens, N., Bond, W.J., Midgley, G. and Lehmann, C.E.R. 2018. Human impacts in African savannas are mediated by plant functional traits. *New Phytologist* **220**: 10–24.

118 McLaughlin, A. and Mineau, P. 1995. The impact of agricultural practices on biodiversity. *Agriculture, Ecosystems and the Environment* **55** (3): 201–212.

119 D'Antonio, C.M. and Vitousek, P.M. 1992. Biological invasions by exotic grasses, the grass/fire cycle, and global change. *Annual Review of Ecological Systematics* **23**: 63–87.

120 Bobbink, R., Hornung, M. and Roelofs, J.G.M. 1998. The effects of air-borne nitrogen pollutants on species diversity in natural and semi-natural European vegetation. *Journal of Ecology* **86**: 717–738.

121 Lindsey, P.A., Balme, G.A., Funston, P.J., Henschel, P.H. and Hunter, L.T.B. 2016. Life after Cecil: Channelling global outrage into funding for conservation in Africa. *Conservation Letters* **9** (4): 296–301.

122 See for example Lewis-Williams, J.D. and Biesele, M. 1978. Eland hunting rituals among northern and southern San groups: Striking similarities. *Africa* **48**: 117–134.

123 Sutherland, L. 18th February 2022. Tanzania, siding with UAE firm, plans to evict Maasai from ancestral lands. *Mongabay*.

124 Ghai, R. 10th June 2022. Fortress conservation: Tension in Tanzania's Loliodo as troops move in to evict Maasai to make way for a game reserve. *Down to Earth, Friday*.

125 Batavia, C., Nelson, M.P., Darimont, C.T., Paquet, P.C., Ripple, W.J. and Wallach, A.D. 2019. The elephant (head) in the room: A critical look at trophy hunting. *Conservation Letters* **12**: e12565. DOI:10.1111/conl.12565.

126 Lindsey, P.A., Balme, G.A., Booth, V.R. and Midlane, N. 2012. The significance of African lions for the financial viability of trophy hunting and the maintenance of wild land. *PLoS One* **7** (1): e29332. DOI:10.1371/journal.pone.0029332.

127 Lindsey, P.A., Miller, J.R.B., Petracca, L.S., Coad, L., Dickman, A.J., et al. 2018. More than $1 billion needed annually to secure Africa's protected areas with lions. *Proceedings of the National Academy of Sciences* **115**: 45.

128 What the Science Says. 2020. Estimating the number and biomass of pheasants in Britain. https://www.whatthesciencesays.org/estimating-the-number-and-biomass-of-pheasants-in-britain/

129 Mason, L.R., Bicknell, J.E., Smart, J. and Peach, W.J. 2020. *The Impacts of Non-Native Gamebird Release in the UK: An Updated Evidence Review*. RSPB Research Report No. 66. RSPB Centre for Conservation Science, Sandy.

130 Dickman, A., Cooney, R., Johnson, P. J., Louis, M. P. and Roe, D. 2019. Trophy hunting bans imperil biodiversity, *Science* **365**: 874.

131 Raine A.F. Gauci, M. and Barbara, N. 2015. Illegal bird hunting in the Maltese Islands: An international perspective. *Oryx*. DOI:10.1017/S0030605315000447

132 Buchan, C., Franco, A.M.A., Catry, I., Gamero, A., Klvaňová, A. and Gilroy, J.J. 2021. Spatially explicit risk mapping reveals direct anthropogenic impacts on migratory birds. *Global Ecology and Biogeography* **31**: 1707–1725.

133 Massei, G., Kindberg, J., Licoppe, A., Gačiç, D., Šprem, N. et al. 2014. Wild boar populations up, numbers of hunters down? A review of trends and implications for Europe. *Pest Management Science* **71**(4): 492–500. DOI:10.1002/ps.3965

134 Turvey, S. 2008. *Witness to Extinction: How We Failed to Save the Yangtze River Dolphin.* Oxford University Press, Oxford.

135 Ellis, E.C., Gauthier, N., Klein Goldewijk, K., Bliege Bird, R., Boivin, N., et al. 2004. People have shaped most of terrestrial nature for at least 12,000 years. *Proceedings of the National Academy of Sciences* **118** (17): e2023483118. DOI:10.1073/pnas.2023483118.

136 Bakker, E.S., Gill, J.L., Johnson, C.N., Vera, F.W.M., Sandon, C.J., et al. 2015. Combining paleo-data and modern exclosure experiments to assess the impact of megafauna extinctions on woody vegetation. *Proceedings of the National Academy of Sciences* **113** (4): 847–855.

137 Habeck, C.W. and Schultz, A.K. 2015. Community-level impacts of white-tailed deer on understorey plants in North American forests: A meta-analysis. *AoB Plants* **7**: plv119.

138 Dublin, H.T. 1995. Vegetation dynamics of the Serengeti-Mara ecosystem: The role of elephants, fire and other factors. In: Sinclair, A.R.E. and Arcese, P. (eds.) 1995. *Serengeti II: Dynamics, Management and Conservation of an Ecosystem.* University of Chicago Press, Chicago, IL and London.

139 Temple, S.A. 1977. Plant-animal mutualism: Co-evolution with dodo leads to near extinction of plant. *Science* **197** (4306): 885–886.

140 Owadally, A.W. 1979. The dodo and the Tambalacoque tree. *Science* **203** (30): 1363–1364.

141 Whitmer, M.C. and Cheke, A.S. 1991. The dodo and the Tambalacoque tree: An obligate mutualism reconsidered. *Oikos* **61** (1): 133–137.

142 Lim, D.Y.H., Starnes, T. and Plumptre, A.J. 2022. Global priorities for biodiversity conservation in the United Kingdom. *Biological Conservation* **277**: 109798.

143 Anon. 2009. The potential environmental and social impacts of a plantation project in Uruguay with tools for planning and monitoring. A Report to Stora Enso.

144 Rodríguez, C., Leoni, E., Lezama, F. and Altesor, A. 2003. Temporal trends in species composition and plant traits in natural grasslands of Uruguay. *Journal of Vegetation Science* **14**: 433–440.

145 WWF. 2014. *The Growth of Soy: Impacts and Solutions.* WWF International, Gland.

146 Carrere, R and Lohmann, L. 1996. *Pulping the South: Industrial Tree Plantations and the World Paper Economy.* Zed Books and The World Rainforest Movement, London and Montevideo.

147 Baguette, M., Deceuninck, B. and Muller, Y. 1994. Effects of spruce afforestation on bird community dynamics in a native, broad-leaved forest area. *Acta Oecologica* **15**: 275–288.

148 Baldi, G., Guerschmann, J.P. and Paruelo, J.M. 2006. Characterizing fragmentation in temperate South America grasslands. *Agriculture, Ecosystems and Environment* **116**: 197–208.

149 Evans, J. and Turnbull, J.W. 2005. *Plantation Forestry in the Tropics: The Role, Silviculture and Use of Planted Forests for Industrial, Social, Environmental and Agroforestry Purposes*, 3rd edition. Oxford University Press, Oxford.

150 Mazzolli, M. 2010. Mosaics of exotic forest plantations and native forests as habitat of pumas. *Environmental Management* **46**: 237–253.

151 See for instance Higgins, J.V., Ricketts, T.H., Parrish, J.D., Dinerstein, E., Powell, G. et al. 2004. Beyond Noah: Saving species is not enough. *Conservation Biology* **18** (6): 1672–1673.

152 https://www.doc.govt.nz/our-work/kakapo-recovery/what-we-do/current-conservation/ (accessed 3/4/2023).

153 Hobbs, R.J., Higgs, E. and Harris, J.A. 2009. Novel ecosystems: Implications for conservation and restoration. *Trends in Ecology and Evolution* **24** (11): 599–605.

154 Marris, E. 2011. *The Rambunctious Garden: Saving Nature in a Post-Wild World*. Bloomsbury Publishing, London.

155 Bánfi, A. 5th February 2022. "Toxic" rewilding rhetoric is "unhelpful and inapplicable to Welsh communities". *Cambrian News*, Aberystwyth, Dyfed UK.

156 Wynne-Jones, S., Strouts, G. and Holmes, G. 2018. Abandoning or reimagining a cultural heartland? Understanding and responding to rewilding conflicts in Wales – The case of Cambrian wildwoods. *Environmental Values* **27**: 377–403.

157 Wroe, S., Field, J., Fullagar, R. and Jermin, L.S. 2004. Megafaunal extinction in the late quaternary and the global overkill hypothesis. *Alcheringa* **28**: 291–331.

158 Jepson, P. and Blythe, C. 2020. *Rewilding: The Radical New Science of Ecological Recovery*. Icon Books, London.

159 Flannery, T.F. 1994. *The Future Eaters: An Ecological History of the Australasian Lands and People*. Reed Books, Chatswood, NSW.

160 Flannery, T. 2001. *The Eternal Frontier: An Ecological History of North America and its Peoples*. Grove Atlantic, New York.

161 Martin, P.S. 2005. *Twilight of the Mammoths: Ice Age Extinctions and the Rewilding of America*. University of California Press, Oakland.

162 Sandom, C., Faurby, S., Sandel, B. and Svenning, J.C. 2014. Global late quaternary extinctions linked to humans, not climate change. *Proceedings of the Royal Society B: Biological Sciences* **281**: 20133254.

163 Araujo, B.B., Oliveira-Santos, L.G.R., Lima-Ribeiro, M.S., Diniz-Filho, J.A.F. and Fernandes, F.A. 2017. Bigger kill than chill: The uneven roles of humans than climate on late Quaternary megafaunal extinctions. *Quaternary International* **431**: 216–222.

164 Ceballos, G., Ehrlich, P.R., Barnosky, A.D., Garcia, A., Pringle, R.M. and Palmer, T.M. 2015. Accelerated human-induced species losses: Entering the sixth mass extinction. *Science Advances* **1**: 1400253.

165 Donlan, C.J., Berger, J., Bock, C.E., Bock, J.H., Burney, D.A. et al. 2006. Pleistocene rewilding: An optimistic agenda for twenty-first century conservation. *The American Naturalist* **168**: 660–681.

166 Perino, A., Perira, H.M., Navarro, L.M., Fernández, N., Bullock, J.M. et al. 2019. Rewilding complex ecosystems. *Science* **364**: eeav5570. DOI:10.1126/science.aav5570.

167 Monbiot, G. 2013. *Feral: Searching for Enchantment on the Frontiers of Rewilding*. Allen Lane, London.

168 Scott, J.C. 2017. *Against the Grain: A Deep History of the Earliest States*. Yale University Press, New Haven, CT and London.

169 Tree, I. 2018. *Wilding*. Picador Books, London.

170 Lever, C. 1992. *They Dined on Eland: The Story of the Acclimatisation Societies*. Quiller Press, London.

171 Dudley, N. 2011. *Authenticity*. Earthscan Books, London.

172 MacKinnon, K.S. 1978. Competition between red and grey squirrels. *Mammal Review* **8** (4): 185–190.

173 Tompkin, D.M., Sainsbury, A.W., Nettleton, P., Buxton, D. and Gurnell, J. 2002. Parapoxvirus causes a deleterious disease in red squirrels associated with UK population declines. *Proceedings of the Royal Society London* **269**: 529–533.

174 Gurnell, J., Wauters, L.A., Lurz, P.W.W. and Tosi, G. 2004. Alien species and interspecific competition: Effects of introduced eastern grey squirrels on red squirrel population dynamics. *Journal of Animal Ecology* **73**: 26–35.

175 https://www.northernredsquirrels.org.uk/squirrels/where-to-see-reds/

176 Twining, J.P., Montgomery, W.I. and Tosh, D.G. 2020. The dynamics of pine marten predation on red and grey squirrels. *Mammalian Biology* **100**: 285–293.

177 Bertolino, S. and Genovesi, P. 2003. Spread and attempted eradication of the grey squirrel (*Sciurus carolinensis*) in Italy, and consequences for the red squirrel (*Sciurus vulgaris*) in Eurasia. *Biological Conservation* **109**: 351–358.

178 Sykes, N. 2017. Fair game: exploring the dynamics, perception and environmental impact of 'surplus' wild foods in England 10kya–present. *World Archaeology* **49** (1): 61–72.

179 Greshko, M. 25th July 2016. New Zealand announces plan to wipe out invasive predators. *National Geographic*.

180 Pearce, F. 2016. *The New Wild: Why Invasive Species Will be Nature's Salvation*. Icon Books, London.

181 Ruland, F. and Jesche, J.M. 2020. How biological invasions affect animal behaviour: A global cross-taxonomic analysis. *Journal of Animal Ecology* **89** (11): 2531–2541.

182 Clarke, G.S., Hudson, C.M. and Shine, R. 2020. Encounters between freshwater crocodiles and invasive cane toads in north-western Australia: Does context determine impacts. *Australian Zoologist* **41** (1): 94–101.

183 Gibbs, G. 2006. *Ghosts of Gondwana: The History of Life in New Zealand*. Craig Potton Publishing, Nelson.

184 Rymer, L. 1976. The history of ethnobotany of bracken. *Botanical Journal of the Linnean Society* **73** (1–3): 151–176.

185 Güereña, D., Neufeldt, H., Berazneva, J. and Duby, S. 2015. Water hyacinth control in Lake Victoria: Transforming an ecological catastrophe into economic, social and environmental benefits. *Sustainable Production and Consumption* **3**: 59–69.

186 Falkenhaug, T. 2014. Review of jellyfish blooms in the mediterranean and black sea. *Marine Biology Research* **10** (10): 1038–1039.

7

RIGHTS IN SYNERGY

We're up in a remote village in northern Vietnam around 2008, talking with an ethnic minority group about a recent ban on hunting highly endangered group of monkeys. It's taken some time to reach them; two day's driving from Hanoi, then switching to motorbikes to navigate some dirt roads and finally on foot up into the mountains. The people are mainly dressed in traditional and beautifully coloured materials and the older women have lacquered their teeth black. There is a tiny generator in the river and some development organisation has installed simple clay-made stoves. The people migrated across the border from China about thirty years previously to avoid political problems. Everyone is cautiously welcoming: we sit down to eat and then start discussions.

The species of monkey is in serious trouble and as conservationists we are supportive of the ban, but also worried about what it will mean to this remote group of people. Hunting is a key part of the fabric of society. Will the men start taking more opium? Could there be increased domestic violence? We are hoping to talk with men and women separately and have a local woman sociologist who is ready to do separate interviews but our hosts announce the discussion will be all together. The men start and give bland assurances that they support the actions of the government. It doesn't mean anything; I suspect that it will be difficult for the women to speak in this setting and start to think we will have wasted the best part of a week. The place is beautiful though and I comfort myself by looking around at the view. Then the women start to talk and suddenly the whole tenor of the conversation changes. They love the ban! Monkeys don't taste good and

DOI: 10.4324/9780429346675-7

they're a hassle to cook. The men just sit around in the forest and get drunk. Now they're home they are actually working on the farms. One woman says that her husband even cleaned out the pigs and all the women go off into peals of laughter. The men look a bit shamefaced and start laughing too. This isn't what we expected at all and we have learned a lot. Not such a wasted trip after all.

The last section has laid out a series of problems; situations where bio-diversity rights come into conflict with other legitimate and important rights. But it doesn't have to be like this, or at least not to an irreconcilable extent. And as the Vietnamese example shows, it is often hard to judge what will happen, defending the rights of other species and of ecosystems can paradoxically support human rights in ways that we might not expect. The people who shout loudest do not necessarily hold the majority view.

While I am deeply suspicious of the concept of "do not harm" because virtually anything has an impact on *someone* or *something*, I am much more convinced by the potential of negotiations and trade-offs, where if no-one gets exactly what they want neither is anyone wholly disadvantaged.[1]

If we take Baird Callicott's view that human rights, animal rights and biodiversity rights are always going to be in tension, what do we do? It is fairly clear that we need to negotiate. Around the turn of the century, when there was first serious discussion about trade-offs,[2] there was a strong reac-tion from some colleagues that the conservation movement should never trade; that the issues we were working on were simply too important. And when issues of human rights started to gain a higher profile in conservation debates the talk was all about "win-win situations", when somehow human wants and biodiversity rights could be magically reconciled. Wherein, I suspect, lie a lot of our problems.

Win-win situations certainly exist. But they are rare and they seldom work for all stakeholders. And, just as important from a practical perspective, they may take some time to become obvious, making the job of persuading people much more difficult. Normally there is give and take and with luck and patience some mutually acceptable compromise. For instance, there is now excellent evidence that setting up a marine protected area or a related conservation initiative such as a Locally Managed Marine Area (LMMA)[3] can increase net fishing catch in the region at the same time as conserving other marine species.[4,5] Providing a safe place for fish to breed and protect-ing a proportion of the largest individuals that are also the most fecund can result in a net increase and spillover into fishing grounds. But it generally takes time to see the increase[6] and conservationists are left with the job of persuading deeply suspicious fishing communities that it is worth a gamble. It works in those places where set asides have been used historically but is still staunchly resisted in other coastal regions where communities are not used to the concepts and assume that this is just conservation by the back door.

The challenges are undeniable. But this is not to say that the situation is hopeless. Over the past few decades, we have learned far more about the practical steps involved in conserving animals and in the last twenty years this has increasingly been integrated with concern for human rights. In the following sections, we'll look first at how concepts of biodiversity rights emerged, then at how they are protected under international and national laws, treaties and conventions, and finally try to pull together some thoughts about how the various questions in the previous section might be answered.

NOTES

1 Maginnis, S., Jackson, W. and Dudley, N. 2004. Conservation landscapes. Whose landscapes? Whose trade-offs? In: McShane, T.O. and Wells, M.P (eds.) *Getting Biodiversity Projects to Work*. Columbia University Press, New York: 321–339.

2 Ibid.

3 Govan, H. 2015. Area-Based Management Tools for Coastal Resources in Fiji, Kiribati, Solomon Islands, Tonga And Vanuatu. Volume 1: Status, capacity and prospects for collaborative resource management. Volume 2: Country reports. Report for the Marine and Coastal Biodiversity Management in Pacific Island Countries (MACBIO) project, Suva, Fiji Islands.

4 Goetze, J.S., Claudet, J., Januchowski-Hartley, F., Langlois, T.J., Wilson, S.K., White, C., Weeks, R., Jupiter, S.D. 2018. Demonstrating multiple benefits from periodically harvested fisheries closures. *Journal of Applied Ecology* **55**: 1102–1113.

5 Roberts, C.M. and Hawkins, J.P. 2000. *Fully Protected Marine Reserves: A Guide*. WWF Endangered Seas Campaign, Washington DC and Environment Department, University of York, York.

6 Ban, N.C., Gurney, G.G., Marshall, N.A., Whitney, C.K., Mills, M., Gelcich, S., Bennett, N.J., Meehan, M.C., Butler, C., Ban, S., Tran, T.C., Cox, M.E., and Breslow, S.J. 2019. Well-being outcomes of marine protected areas. *Nature Sustainability* **2**: s24–s32.

8

WHAT DO OTHERS SAY?

A group of about 20 of us are sitting around a long, oblong table in Thimphu, Bhutan. A Buddhist elder is making a formal introduction to the meeting and we are all waiting to take a sip from the bowl of liquid in front of us, a broth with a thick scum of butter on top. Suddenly, at the height of the ceremony, one of the monks darts forward and takes a photograph; there's a huge flash and everyone jumps. I love those times when modernity bursts into an ancient ritual; like a monk taking a call on his mobile phone in the middle of a meeting or the priest who winked at me and said hello in the middle of a very formal church service when I was still a tiny boy. The meeting in Bhutan was the scene of a more significant juxtaposition as well. I'd been working with Liza Zogib and others to bring together religious scholars and conservationists from around the Eastern Himalayas to search for common ground in the conservation of this unique and beautiful part of the world. We had representatives from Buddhist, Bon and animist traditions and a disparate group of conservation scientists from the region and beyond, some of whom were also practising Buddhists. Over a week we explored matters of mutual concern. We were talking in dramatically different languages: conservation corridors and protected areas on the one hand, hierarchies of gods and sacred mountains on the other. But gratifyingly, albeit not particularly surprisingly, our prescriptions for the future were remarkably similar, it's just that we were coming at the problem from a different perspective.[1]

DOI: 10.4324/9780429346675-8

THE ROLE OF RELIGIOUS FAITHS

The large majority of the world's population believe in and to some extent follow a religious doctrine; most people follow one of the major, organised religions, a significant minority believe in local traditions and gods and more people than you might expect believe in a mixture of the two. And research shows that the strength of religious belief is concentrated in areas of the highest biodiversity.[2] So any examination of what other say can usefully start with the attitudes of the world's great religions, although it certainly shouldn't end there.

As noted earlier, faiths originating in Central and South Asia, China and Japan (Buddhism, Daoism, Hinduism, Jainism, Shinto, Sikhism and Zoroastrianism) tend to view nature as a critical aspect of the divinity.[3] In contrast, the three monotheistic faiths that came from the Middle East (Christianity, Judaism and Islam) have strong teaching against idolatry and reject the concept of sacred species or sacred sites. Sacred groves were sometimes destroyed by Christian missionaries because of their identification with idolatry, to destroy sacred competitors, or even as a punishment.[4] Maverick dissidents from this view, like Giovanni di Pietro di Bernardone, better known as St Francis of Assisi and the founder of the Franciscan order of Christian monks, were often venerated in theory but their ideas were not followed through in practice. In an article in *Science* in 1967 Lynn White, a Christian, argued that Christians, Jews, and Muslims have abused the environment because it is viewed as God's creation to serve Humankind.[5] This created a reaction, particularly among Christians, who were the main targets of White's critique, and there have since been strong statements of support for the environment from Christian leaders, e.g., Pope John Paul II[6] and Pope Benedict XVI.[7]

Interest in the potential role of religions as a way of building support for ecosystem rights has been increasing.[8] The ecological attitudes of most of the world's faiths have now been thoroughly examined, including Buddhism,[9] Taoism,[10] Hinduism,[11] Jainism,[12] Judaism[13] and Islam.[14] The differences between Eastern and Western faith groups are significant but should not be overstated; there has been much environmental damage in countries where faiths revere nature as an aspect of the sacred. Conversely, John Muir, the American traveller and naturalist who is regarded by many as the founder of the modern environmental movement, was driven by a deep and passionate Christianity. Monotheistic faiths have sometimes advocated forms of land management like those promoted by conservationists, as in the *hima* system, a land protection system originating in the Arabian Peninsula some 1,400 years ago and adopted by many Islamic societies.[15]

Note that many of the analyses quoted above are general statements about the environment, by no means all address biodiversity rights or indeed the rights of nature in general. Some of the largest contributions comes almost accidentally from faith groups and often from minor faiths in terms of number of followers. Sacred natural sites, sites set aside for religious purposes, can have very high levels of biodiversity because they are strictly protected and well-guarded,[16,17] but nature conservation was not the original aim. Additionally, and more deliberately, a growing number of religious institutions across a variety of faiths see care for wild species or environments as part of their practice. This can range from monks caring for particular species,[18] through to management of land for conservation,[19] including land around religious buildings. In parts of Ethiopia, for instance, the grounds of Christian orthodox churches are now the main refuge of native forests.[20] Another important way in which faiths influence this debate is through their teaching, and the opportunities to influence followers. Additionally, the question of how religions invest their money is gaining increasing attention; the net worth of the world's religions is measured in trillions of dollars and a clear commitment to avoid investments that damage nature would be a considerable game changer.[21]

MOVES TOWARDS A RECOGNITIONOF BIODIVERSITY RIGHTS

Much has already been written about the history of the environmental movement and the development of a philosophy of conservation, I will pick only a few highlights here. Aldo Leopold died in 1948, fighting a grass fire on a neighbour's ranch and shortly after being appointed an advisor on conservation to the United Nations. *A Sand County Almanac* was published posthumously and is both a nature journal of a year in rural Wisconsin and a series of essays and sketches. In it, Leopold outlined a land ethic: *a thing is right when it tends to preserve the integrity, stability and beauty of the biotic environment. It is wrong when it tends otherwise.*[22] Conservationists today might question the word "stability" when we know so much more about the extent of natural flux in ecosystems but Leopold's ethic stands up surprisingly well after 80 years or so.

A bucket of cold water was thrown over the young environmental movement in 1979 by Norman Myers in his book *A Sinking Ark*,[23] which argued persuasively that conservation was failing, and in particular pointed to the rapid loss of the world's tropical forests. My copy is well thumbed and marked up. I'm not sure I even finished it but it persuaded me to stop what I was doing and spend the next years working on tropical forest conservation. Note though that Myers didn't tackle rights as such, I suspect he was, like many of us, acting on an instinctive level that a human-induced species crash was simply wrong. In later writing, he stressed the practical implications of this in terms of lost resources and opportunities.[24] The bulk of attention since then, including our own, has been focused on the benefits from biodiversity rather than on the ethical issues, although most people I know working in the conservation field are driven by far more than simply utilitarian considerations.

Meanwhile, the Gaia concept, the idea that the whole planet acts like a single living entity,[25] and some other examples of big picture thinking[26,27] were gradually leading to changes in attitude. These ideas stressed the need for coexistence and the pragmatic reasons for conserving biodiversity within an emerging ethical framework. It is sometimes still a little difficult to see where one starts and the other stops. The concept of Earth jurisprudence was first identified by Thomas Berry, Catholic priest, scholar and writer on world religions. He argued for law and human governance based on recognition that humans are part of a wider community of living beings and that the welfare of each is dependent on the welfare of the Earth as a whole. In an important statement he argued, amongst other things, that *The universe is a communion of subjects not a collection of objects* and that *every component of the Earth Community has three rights: The right to be, the right to habitat, and the right to fulfil its role in the ever-renewing processes of the Earth Community.* And importantly, *Human rights do not cancel out the rights of other modes of being to exist in their natural state.*[28]

Ideas were discussed at a conference organised by the Gaia Foundation, and first articulated in detail by Cormac Cullinan in his book *Wild Law* in 2002.[29] He points out that extinction of a species or an ecosystem is not even a crime in most countries (there are now some moves to address this) and argues that while the interpretation of Earth jurisprudence will vary from country to country, it should contain some common elements, summarised below:

- A recognition that the source of such rights is the universe rather than human governance systems.
- Recognition of the roles of non-humans and restraint of humans from unjustifiably preventing them from fulfilling these roles.

- Concern for reciprocity and maintenance of dynamic equilibrium between all members of the Earth Community, determined by what is best for the whole system (defined as Earth justice).
- A practice of condoning or disapproving human conduct depending on its impacts on the Earth Community.[30]

Recognising the whole universe as the source of rights is an approach that certainly can't be criticised for thinking too narrowly. Cullinan argues to *rediscover Earth jurisprudence...it is essential to start by looking at the fundamental laws and principles of the universe, since these provide the ultimate framework within which any human legal framework must exist.*[31] The concept has been gaining ground, albeit still within a subsector of the conservation community, and there have been several follow-up workshops and research arguing for these issues to be brought more centrally into approaches to land and water management.[32]

In 2009, the United Nations General Assembly proclaimed 22 April as International Mother Earth Day. In so doing, Member States

acknowledged that the Earth and its ecosystems are our common home, and expressed their conviction that it is necessary to promote Harmony with Nature in order to achieve a just balance among the economic, social and environmental needs of present and future generations.[33]

A series of statements have emerged since from the UN on the need for humans to live in balance with the natural world. There have also been by 2023 11 interactive dialogues, one a year, covering issues including reconnecting with nature, consumption, Indigenous knowledge, Earth Jurisprudence and alternatives to Gross Domestic Product as a measure of wellbeing. Dialogue four examined the issue of nature as an equal partner to humankind, which is probably the closest match to the issues discussed here.

A handful of countries and individuals have been steadily pushing this agenda. In 2010, Bolivia hosted a large meeting of climate justice activists in the wake of the United Nations Copenhagen climate summit, which was universally recognised as a failure. (Frustration at this failure also paved the way for the more successful Paris climate meeting some years later.) The World People's Conference on Climate Change and the Rights of Mother Earth took place in Cochabamba, which amongst other results published a manifesto: *Rights of Nature:The universal declaration of the rights of Mother Earth*.[34] This declaration was considerably more radical in content and intent than simply looking at rights of nature or biodiversity rights, noting that *Mother Earth is a living being* and recognising *organic and inorganic beings*, as well as nailing the capitalist system as the major cause of environmental problems, thus probably ensuring that many stakeholders would hold back from full support. It also stressed some important issues that we have been discussing here, identifying a *living community of interrelated and interdependent beings with a common destiny* and the *right of life to exist* as well as the important caveat that *the rights of each being are limited by the rights of other beings and any conflict between their rights must be resolved in a way that maintains the integrity, balance and health of Mother Earth.*[35]

Over the last decade, a plethora of different stakeholder groups have been directly addressing nature's rights. A United Nations website[36] lists policy initiatives from over 40 countries, some with multiple actions. The list is eclectic, ranging from UN bodies to individual town councils and embracing NGOs, Indigenous peoples, religious bodies, universities, sporting events, the European Parliament and national governments. I rather like this hodgepodge of stakeholders because it shows that the ideas are gaining ground at every level and it is clear that the UN list is far from complete.

For example, a group of children in Argentina filed a Collective Environmental Appeal to the Supreme Court of Justice to protect the Paraná Delta, including "recognition of rights". In Brazil, a legal proposal for the Rights of Nature was presented to the Sao Paulo Commission on Legality and is expected to gain further support. A citizens' petition on Rights of Nature was presented to the Senate in Chile. In the United States, conservation and rights groups are drafting a Universal Declaration of River Rights. Rights for particular places have been recognised or lobbied for in amongst others El Salvador, France, Hungary, New Zealand, Sweden and Switzerland. The town council of Civita Castellana in Italy became the first in Europe to declare the municipality a "nature rights zone". References to Harmony of Nature have started to creep into United Nations business, such as in the UN Forum on Forests.[37] The Ecuadorian constitution and the Colombia Supreme Court both recognise the rights of nature, in the latter case for the Amazon. And in India the Uttarakhand High Court recognised that the Animal Kingdom has the rights, duties and liabilities of a living person.[38]

Most recently, the United Nations General Assembly published a new resolution on "harmony with nature",[39] which at first sight appears to address many of the issues discussed here. This is an important statement, and I am delighted to see it adopted, but careful reading suggests it still falls short of recognising fundamental global rights for biodiversity. Text refers to *coexistence of humankind in harmony with nature* and the *symbiotic connection between human beings and nature that fosters a mutually beneficial relationship,* to nature-based solutions and the problems facing humanity due to our mismanagement of nature. All good and necessary stuff. But the nearest to the focus of this book is a note that *...some countries recognize the rights of nature or Mother Earth in the context of the promotion of sustainable development.* This has been stated before, in the outcome document of the United Nations Conference on Sustainable Development in 2012.[40] Maybe I am being too picky, but this reads to me like less than a total acceptance of this as a concept. Particularly as every other "right" mentioned in the lengthy document, rights to education, food, exploitation of minerals, etc., was simply a "right" and not something qualified by saying it is only recognised by a few countries. And this in a statement that is specifically about the environment at a meeting recognising 20 years after the Earth Summit set up the three so-called Rio Conventions addressing biodiversity, climate change and desertification. This ambivalence plays out in the way that biodiversity rights have been addressed in practice, as will become clear in the following chapter.

NOTES

1 Higgins-Zogib, L., Dudley, N. and Aziz, T. (eds.) 2012. *The High Ground: Biocultural Diversity and Conservation of Sacred Natural Sites in the Eastern Himalayas.* WWF Bhutan, Thimphu.

2 Bhagwat, S.A., Dudley, N. and Harrop, S.R. 2012. Religious following in biodiversity hotspots: Challenges and opportunities for conservation and development. *Conservation Letters* **4**: 234–240.

3 Nasr, S.H. 1996. *Religion and the Order of Nature: The 1994 Cadbury Lectures at the University of Birmingham.* Oxford University Press, Oxford.

4 Adler, J. 2006. Cultivating wilderness: Environmentalism and legacies of early Christian asceticism. *Comparative Studies of Society and History* **48**: 4–37.

5 White, L. 1967. Historical roots of our ecologic crisis. *Science* **155**: 1203–1207.

6 Pope John Paul II. 1990. The ecological crisis a common responsibility peace with God The Creator, peace with all of creation. Message of His Holiness for the celebration of the World Day of Peace 1 January. Catholic Conservation Center, Wading Bridge, New York.

7 Pope Benedict XVI. 17th July 2008. Address by his Holiness Benedict XVI. Bangaroo, Sydney, Australia. Holy See, Vatican City.

8 Tucker, M.E. and Berling, J. 2003. *Worldly Wonder: Religions Enter Their Ecological Phase.* Open Court Publishing, Chicago, IL.

9 Tucker, M.E. and Williams, D.R. (eds.) 1998. *Buddhism and Ecology: The Interconnection of Dharma and Deeds.* Harvard University Press, Cambridge, MA.

10 Snyder, S. 2006. Chinese traditions and ecology: Survey article. *Worldviews, Environment, Culture and Religion* **10**: 100–134.

11 Narayanan, V. 2001. *Water, Wood, and Wisdom: Ecological Perspectives from the Hindu Traditions.* Daedalus, American Academy of Arts and Science, Cambridge, MA.

12 Singhvi, L.M. 1990. *The Jain Declaration on Nature.* Jainism Global Resource Center, Alpharetta, GA.

13 Vogel, D. 1999. *How Green Is Judaism?* University of California, Berkeley, CA.

14 Bagader, A.A., Al-Chirazi El-Sabbagh, A.T., As-Sayyid Al-Glayand, M. and Izzi-Deen Samarrai, M.Y. 1994. *Environmental Protection in Islam. Environmental Policy and Law Paper 20.* IUCN, Gland.

15 Kilani, H., Serhal, A. and Llewellyn, O. 2007. *Al-Hima: A Way of Life.* IUCN West Asia Regional Office, Amman, Jordan, and the Society for the Protection of Nature in Lebanon, Beirut.

16 Dudley, N., Bhagwat, S., Higgins-Zogib, L., Lassen, B., Verschuuren, B. and Wild, R. 2010. Conservation of biodiversity in sacred natural sites in Asia and Africa: A review of the scientific literature. In: Verschuuren, B., Wild, R., McNeely, J. and Oviedo, G. (eds.) *Sacred Natural Sites: Conserving Nature and Culture.* Earthscan, London: 19–32.

17 Boraiah, K.T., Vasudeva, R., Bhagwat, S.A. and Kushalappa, C.G. 2003. Do informally managed sacred groves have higher richness and regeneration of medicinal plants than state-managed reserve forests? *Current Science* **84**: 804–808.

18 Li, J., Wang, D., Yin, H., Zhaxi, D., Jiagong, Z., Schaller, G.B. et al. 2014. Role of Tibetan Buddhist monasteries in snow leopard conservation. *Conservation Biology* **28**: 87–94.

19 Mallarach, J.M. and Papayannis, T. (eds.) 2007. *Protected Areas and Spirituality.* IUCN and Publicacions de l'Abadia de Montserrat, Gland.

20 Wassie, A., Sterck, F., Teketay, L. and Bongers, F. 2009. Effects of livestock exclusion on tree regeneration in church forests of Ethiopia. *Forest Ecology and Management* **257**: 765–772.

21 Dudley, N., L Higgins-Zogib, and S. Mansourian. 2006. *Beyond Belief: Linking Faiths and Protected Areas for Biodiversity Conservation.* WWF and Alliance of Religions and Conservation, Gland, and Bath.

22 Leopold, A. 1949. *A Sand County Almanac and Sketches Here and There.* Oxford University Press, New York.

23 Myers, N. 1979. *The Sinking Ark: A New Look at the Problem of Disappearing Species.* Pergamon Press, Oxford.

24 Myers, N. 1985. *The Primary Source: Tropical Forests and Our Future.* W.W. Norton and Co., New York.

25 Lovelock, J. 1979. *Gaia: A New Look at Life on Earth.* Oxford University Press, Oxford.

26 Chivian, E. and Bernstein, A. (eds.) 2008. *Sustaining Life: How Human Health Depends on Biodiversity.* Oxford University Press, Oxford.

27 Naess, A. 2008. *Ecology of Wisdom.* Penguin Books, London.

28 Presented at a meeting organised by the Gaia Foundation and reprinted in Cullinan, C. 2011. *Wild Law: A Manifesto for Earth Justice* (2nd edition). Green Books, Cambridge: 103.

29 Ibid.

30 Ibid., page 117.

31 Ibid., page 78.

32 Rühs, N. and Jones, A. 2016. The implementation of Earth Jurisprudence through substantive constitutional rights of nature. *Sustainability* **8** (2): 174. DOI:10.3390/su8020174

33 http://www.harmonywithnatureun.org/

34 Morales Ayma, E., Barlow, M., Bassey, N., Biggs, S., Cullinan, C. et al. 2011. *The Rights of Nature: The Case for a Universal Declaration of the Rights of Mother Earth*. The Council of Canadians, Fundación Pachamama and Global Exchange, San Francisco, CA.

35 Ibid.

36 http://www.harmonywithnatureun.org/rightsOfNaturePolicies/

37 All references from http://www.harmonywithnatureun.org/rightsOfNaturePolicies/

38 Chapron, G., Epstein, Y. and López-Bao, J.V. 2019. A rights revolution for nature. *Science* **363** (6434): 1392–1394.

39 United Nations. 2020. Resolution adopted by the General Assembly on 21 December 2020. 75/220. Harmony with Nature. 75th Session, agenda item 19 (g).

40 United Nations. 20–22 June 2012. *The Future We Want*. Outcome document of the United Nations Conference on Sustainable Development, Rio de Janeiro.

9

CURRENT PROTECTION
OF BIODIVERSITY RIGHTS

Around 12,000 people had gathered in Montreal in December 2022 for a debate about what many saw as the future of life on the planet. The 15th Conference of Parties of the UN Convention on Biological Diversity had been billed as the "Paris moment" for biodiversity, harking back to the 2015 meeting in Paris where meaningful targets to reduce climate change were agreed for the first time. The build-up to the CBD meeting had already been fraught. Originally scheduled for Kunming in China in 2020, the gathering had been delayed several times, ostensibly due to the Covid-19 pandemic, but also by a continuing failure to reach agreement. Some highly ambitious targets were being proposed, including the protection and conservation of almost a third of the planet's land, freshwater and marine environment. Successive planning meetings had ground to a halt with much of the text still in debate. The first two weeks in Montreal saw glacially slow progress, with governments holding fixed positions and working groups abandoning their attempts to reach consensus, there were walkouts, briefings and counter-briefings and a general feeling of malaise. Those of us at the meeting as representatives of non-governmental organisations were not allowed to speak in official sessions, so we sat at the back watching while the same arguments kept circling time and again. Almost at the last minute, the Chinese chair proposed a consensus text and – at four o'clock in the morning – exhausted delegates signed off. But even then the drama wasn't over because the Democratic Republic of Congo (DRC) raised an objection and said they would not abide with the decision – CBD decisions require complete consensus – and a further 24 hours went by before the DRC was satisfied and the meeting was brought to an end.

DOI: 10.4324/9780429346675-9

The resulting Global Biodiversity Framework (GBF)[1] is, as had been hoped, the most ambitious plan to save biodiversity to date, although many NGOs still criticised individual aspects[2] and there are targets that remain so general that they are unlikely to result in meaningful action. But compared with what might have been expected even a decade ago, this represents an extraordinary change in the attitudes of governments and civil society. This is going to be the piece of international legislation for many conservation groups over the coming decade. But what does it say, if anything, about rights?

This chapter is not primarily about legislation and policies for the protection of biodiversity. There is already a huge array of these available at international, regional, national and local level, applied with varying degrees of seriousness. In line with the central question addressed by this book, the following discussion is instead about whether these policies acknowledge the rights of biodiversity – or rather the rights of biodiversity apart from humans – per se, or whether biodiversity is only valued for utilitarian or at least human-centred purposes.

Despite the growing interest in the rights of nature and harmony with nature, quoted in the last section, it has proven surprisingly hard to find other more specific language in laws or treaties that could be interpreted as relating to biodiversity rights, and not at all easy to find references to values. In fact, the trend currently seems to be running in the opposite direction.

The CBD itself has in practice always stressed the value that humans receive from biodiversity more than the intrinsic value of biodiversity, or any rights that biodiversity might have itself. Initial discussions about the Convention focused more on issues of intellectual property rights relating to biodiversity, driven by anger amongst many developing nations that the world's wealthy nations were expropriating valuable species from poorer nations and gaining the majority of economic benefits in consequence. Calestous Juma, a Secretary General of the CBD wrote several books on this issue,[3] and much of the early debate at Conferences of Parties centred around the rights of countries to own their genetic resources. Important though these issues are, they are separate from conservation and for the first few years many conservation NGOs saw little reason to engage with the Convention.

This was not the intention of the people who first suggested the need for a convention. Right at the beginning of the Convention text, signed in 1992 at the Earth Summit in Rio de Janeiro, there is clear language recognising the value of biodiversity beyond any uses that we might make of other species and ecosystems: *Conscious of the* **intrinsic value of biological diversity** *and of the ecological, genetic, social, economic, scientific, educational, cultural, recreational and aesthetic values of biological diversity and its components*… (my emphasis). From the perspective taken here, this is an important caveat and a rare early signal from the global community that biodiversity might have rights of its own.

In the years since, and in light of growing recognition of the importance of biodiversity, the objectives of biodiversity conservation have become more ambitious and more explicit. But to an even greater extent than before, the importance of biodiversity is often – in fact almost always – presented through a human lens, my emphasis in all the following quotations. The GBF, signed in December 2022 and described in the opening paragraphs, notes:

> **Biodiversity is fundamental to human well-being** and a healthy planet, and economic prosperity for all people. including for living well in balance and **in harmony with Mother Earth**, we depend on it for food, medicine, energy, clean air and water, security from natural disasters as well as recreation and cultural inspiration, and it supports all systems of life on earth.

Nowhere in the agreement is there language suggesting that biodiversity should be conserved because it has a right to exist or because of its intrinsic values, beyond the rather ambivalent reference to Mother Earth. It could be argued that this is unnecessary given reference in the overall convention text, but so are human benefits from biodiversity and reference to these is made repeatedly in the GBF.

In the last 30 years, the CBD has quite rightly put more emphasis on human rights including the rights of Indigenous peoples and local communities, gender and equity: all very necessary changes. But it is now almost as if signatory states to the Convention have become afraid of acknowledging that biodiversity has any rights other than the right to supply us with things that we need.

Do other global treaties take a different approach? Not in most cases as far as I can see. The Ramsar Convention or International Convention on Wetlands begins by recognising *the interdependence of Man and his environment* before noting that **wetlands constitute a resource** *of great economic, cultural, scientific, and recreational value* and waterfowl *should be regarded as an international resource.*[4] I've worked with the Ramsar Convention and don't doubt for a moment that many of those involved have a powerful sense of the intrinsic value of wetlands, but this doesn't show up in key documents. Similarly, the Convention on Migratory Species starts by RECOGNIZING that wild animals in their innumerable forms are an irreplaceable part of the Earth's natural system which must be conserved **for the good of mankind**.[5] The Convention on International Trade in Endangered Species of Wild Flora and Fauna (CITES) has a more sympathetic approach that nonetheless remains ambiguous: *Recognizing that wild fauna and flora in their many beautiful and varied forms are an irreplaceable part of the natural systems of the earth which* **must be protected for this and the generations to come**.[6] This doesn't come out and say that such species are intrinsically valuable, but it does stress their beauty and irreplaceable nature, nor does it refer narrowly to their use as ecosystem services. Nor is it explicit whether "generations to come" refers only to human generation or all species.

Individual countries have tended to have a more focused approach than international organisations in many respects, illustrating perhaps the widely different views around the world and lack of consensus at global level. Legal initiatives fall into three main groups; explicit recognition of the rights of nature in general within national laws, petitions for the same and – an interesting phenomenon – an increasing tendency to confer legal rights on a particular ecosystem or habitat. A small collection of countries already recognise the rights of nature in a legal sense. A statement from Uganda, as part of the 2019 National Environment Act, is one of the most interesting, recognising *nature's fundamental rights to* **be, evolve and regenerate** (my emphasis), which fits neatly into the definition of biodiversity rights being floated here. Citizens in countries like Portugal and Switzerland are petitioning their governments about rights of nature. Bolivia called for an Earth Assembly, which would recognise a *non-anthropocentric and cosmobiocentric vision*. Conferring rights on a particular place is also a growing phenomenon, although here it feels a little as if concerned stakeholders are drawing on concepts of nature's rights to win particular conservation victories rather than, as yet, achieving an overall acceptance of rights in general. Amongst the places that have or are being supported to have legal rights are rivers such as the St Lawrence in Canada and the Ganges in India (rivers are still the commonest ecosystem to be subject of rights legislation)[7]; island such as Isla de Salamanca in Colombia; Lake Total in Colombia, mangroves in Ecuador, the whole of Mother Nature in India; the Wadden Sea offshore of the Netherlands; and Mar Menor lagoon in Spain. In Australia the Great Ocean Road, a spectacular route south of Melbourne along the coast of Victoria, has been recognised as *one living and integrated natural entity*. Most of these legal changes have occurred in the last few years.[8] There is currently a petition circulating calling for a more general recognition of the rights of rivers.[9] The extent to which these various initiatives have teeth, in terms of providing genuine protection by law against degradation, is still too early to say.

The strongest statements of support for intrinsic values of nature might be expected to come from the non-governmental organisations dedicated to nature conservation. I looked at what some of the largest organisations are saying. The results are variable. Some have mission statements that present their aims without giving a clear explanation of why. My emphasis throughout the following paragraph. WWF International aims to

> build a future in which **people live in harmony with nature**... to conserve biodiversity, the web that supports all life on Earth; reduce humanity's ecological footprint; and ensure the sustainable use of natural resources to support current and future generations.[10]

IUCN, the International Union for Conservation of Nature, aims *to influence, encourage and assist societies throughout the world to* **conserve the integrity and diversity of nature** *and to ensure that any use of natural resources is equitable and ecologically sustainable.*[11] Birdlife International aims *to conserve birds, their habitats and global biodiversity,* **working with people toward sustainability** *in the use of natural resources.*[12] The Wildlife Conservation Society **saves wildlife and wild places** *worldwide through science, conservation action, education, and inspiring people to value nature.*[13] Flora and Fauna International aims **to conserve threatened species and ecosystems** *worldwide.*[14] These statements are all admirably concise and I don't have any disagreement with them but they also don't give any hint as to why this might be necessary. All those organisations have published extensively on the justification for conservation, but even a careful reader will struggle to find many references to intrinsic values and the right to exist.

Some definitely see the distinction. The Nature Conservancy, now probably the largest conservation NGO in the world, clearly recognises the dichotomy and reflects both intrinsic and human values:

> to conserve the lands and waters on which all life depends. Our vision is a world where the diversity of life thrives, and people act **to conserve nature for its own sake** and its ability to fulfil our needs and enrich our lives.[15]

But others have gone in a reverse direction. Conservation International, at one time viewed as the most nature-centric of all the large conservation

NGOs rewrote its mission statement a few years ago to support societies to responsibly and sustainably care for nature **for the well-being of humanity**. It should be noted however that the CI vision is a healthy, prosperous world in which societies are forever committed to caring for and valuing nature, for the long-term benefit of people **and all life on earth**, so perhaps the change is not as fundamental as all that.[16]

I've quoted at length here because organisations spend a long time on their mission statements and they generally say something important about the way the individuals working within them view the world. I don't think the lack of reference to the rights or even the intrinsic value of biodiversity is accidental. Over the past two decades, conservation bodies in both the governmental and non-governmental sectors have come under increasing criticism for ignoring or downplaying human rights in the defence of wildlife. There are some uncomfortable truths in these criticisms. The various conservation stakeholders have been scrambling to address – and to be seen to address – these issues since the turn of the century. What to me seems to be a perverse result of this is a nervousness of talking about the intrinsic values of nature, let alone the rights of nature, for fear of implying that this indicates lack of interest in human rights. No one wants to be a bunny-hugger. Yet the large majority of people who have criticised conservationists for human rights abuses are not anti-nature. Quite the reverse, some of the strongest critics I know are passionate naturalists. I think many conservationists are missing the point.

NOTES

1 UN Environment Programme. 18 December 2022. Kunming-Montreal Global biodiversity framework. Draft decision submitted by the President. Convention on Biological Diversity. Conference of Parties to the CBD, 15th Meeting – Part II, Agenda item 9(a). CBD/COP/15/L.25.

2 https://wwf.panda.org/wwf_news/press_releases/?7334466/WWF-reaction-to-new-COP15-text-published-18-Dec

3 Juma, C. 1989. *The Gene Hunters: Biotechnology and the Scramble for Seeds*. Zed Press, London and Princeton University Press, Princeton, NJ.

4 Convention on Wetlands of International Importance especially as waterfowl habitat, Ramsar, Iran, 2.2.1971, as amended by the Protocol of 3.12.1982 and the Amendments of 28.5.1987. Paris, 13 July 1994.

5 Convention text, Convention on Migratory Species. https://www.cms.int/en/convention-text

6 Convention on International Trade of Endangered Species of Wild Flora and Fauna, Signed at Washington, D.C., on 3 March 1973, Amended at Bonn, on 22 June 1979, Amended at Gaborone, on 30 April 1983 https://cites.org/eng/disc/text.php#II

7 International Water Resources Association. 2019. *Should Rivers Have Rights? Water International Policy Brief Number 13.* IWRI, Paris.

8 All references in this section are from http://www.harmonywithnatureun.org/rightsOfNature/

9 https://www.rightsofrivers.org/

10 https://www.wwf.org.uk/jobs/our-values#:~:text=Our%20mission%20is%20to%20build,support%20current%20and%20future%20generations

11 https://www.iucn.org/about-iucn

12 https://www.birdlife.org/who-we-are/

13 https://www.wcs.org/about-us

14 https://www.fauna-flora.org/

15 https://www.nature.org/en-us/about-us/who-we-are/our-mission-vision-and-values/

16 https://www.conservation.org/about

10

IS MY WAY THE BEST WAY?

I am increasingly suspicious of certainty. One of the things that has struck me while researching this book is just how polarised and angry the debate is, full of people convinced that they have the right answer and brutally dismissive of dissent. I have a huge respect for some of the animal rights activists, like the people campaigning against illegal badger baiting and pit bull fighting who live undercover, constantly threatened by beatings or worse from the organised criminal gangs who make money from deliberately cruel sports. And for the thousand or more protected area rangers who have died in the line of duty in the last decade, while many others have been injured, threatened and their families put at risk.[1] And for the similar, probably larger, numbers of Indigenous peoples and local communities who have suffered and died defending their traditional lands against mining, agriculture and illegal poaching.[2]

I'm also appalled by the flip side, the death threats against anyone working with animals in medical research,[3] or anyone who defends trophy hunting, and I'm slightly bemused by the weird rituals played out in Britain between the hunt saboteurs and the hunters. Similarly, I am frustrated by conservationists who claim the whole conservation enterprise is simply one aspect of sustainable development. There's a real risk of losing focus. When people first started thinking about the impact of the internet I saw few predictions about just how much bile and anger would come to the surface.

This is dangerous. The whole concept of biodiversity conservation is only a few decades old, and in many respects we are still in the earliest stages of understanding how we can co-exist successfully with other species in

DOI: 10.4324/9780429346675-10

the context of a far larger human population, climate change and multiple, rapidly changing ecosystems. The fact that management worked in the past does not necessarily mean it will work in the changing environmental conditions of the present and future. The conservation movement can be as guilty of black and white thinking as anyone else and that has also led us down some blind alleys. Demonising bushmeat hunters, loggers, illegal miners and landless settlers, amongst many others, has encouraged a combative form of conservation that has created serious backlash (and incidentally hasn't worked very well). It is hard to get the balance right; we need a bit of fire in our bellies to stand up against what are often entrenched and powerful interests, but on the other hand righteous anger can all to easily drive us in the wrong direction.

So, this is a plea for tolerance. Ethics are evolving. As we've discussed, balancing human rights, animal rights and biodiversity rights is going to need considerable give and take. And people start from different places depending on their culture, background, religious faith and the experiences in their own lives. We are far less logical than we like to think. And yet the moral issues are real. Conservation organisations have changed their policies over the years, towards hunting by Indigenous people for instance,[4] the management of protected areas, the culling of elephants and many more. The changes will keep coming as we learn more and understand more. We need to keep our eyes and ears open and be ready to learn and adapt as necessary. Hopefully distinguishing biodiversity rights from animal rights and human rights, and understanding where these may be in conflict, can help us avoid some of the problems of the past.

NOTES

1   Galliers, C., Cole, R., Singh, R., Ohlfs, J., Aisha, H. et al. 2022. Conservation casualties: An analysis of on-duty ranger fatalities (2006–2021). *PARKS* **28** (1): 39–50.

2   https://www.globalwitness.org/en/campaigns/environmental-activists/decade-defiance/

3   Hadley, J. 2009. Animal rights extremism and the terrorism question. *Journal of Social Philosophy* **40** (3): 363–378.

4   Burke, D.C. 2021. The case for a Greenpeace apology to Newfoundland and Labrador. *The Northern Review* **51**: 173–187.

11

WHAT SHOULD OUR ROLE BE NOW?

Last summer I returned to the Kinabatangan River in Sabah, Borneo, after a gap of 20 years. The Kinabatangan presents a classic twenty-first century conservation challenge; a strip of mainly natural vegetation running through land that has largely been converted to oil palm. Ignoring state conservation laws, some of the plantations extend right down to the water's edge. Resident orangutan, many birds and the traditional human communities are all increasingly squeezed into smaller and smaller areas. Some forests have been preserved as habitat for the swiftlets whose nests are collected for the extraordinarily valuable birds' nest soup, basically bird spit and mud, which has high value in the Asian food markets. I've visited the caves where men climb up perilous bamboo ladders to collect their harvest, walking over metre-deep droppings heaving with giant cockroaches. (Our knowledge of zoonotic diseases would suggest that is something not to be done as casually as we used to.) A herd of elephants make a perilous, twice-yearly journey along the river, always in danger of being shot by workers in the plantations; I've seen them take to the water and swim to avoid danger areas. My feelings about visiting after a long absence were mixed. Most species still seemed to be hanging on; there were still proboscis monkeys and orangutan and a spectacular array of birds. On my penultimate visit we had been setting up a forest restoration project; it seemed like that had worked but there had been other losses since, including a massive influx of invasive plants that had virtually clogged some of the riverside pools, and attempts to control illegal clearing by individuals and companies were at best partially successful. But the whole ecosystem was still bursting with life; the riot of insect noises

DOI: 10.4324/9780429346675-11

at night contrasted dramatically with the almost total silence in the capital, Kota Kinabalu, where we could sit out by the water in the evening and not see a single insect, poisoned into oblivion by malaria controls. There were still dedicated individuals running projects; we stayed in a simple ecolodge that was surviving after weathering serious losses during the Covid-19 pandemic. As was the case 20 years back, the whole ecosystem is under pressure but still just about hanging on and there is everything to play for.

I've been doing this stuff long enough to have seen many projects go through various stages and depending on my mood I can feel either optimistic or pessimistic. There have certainly been some spectacular gains but also many losses and − most commonly of all − places that simply jog along, seeing things get better in one area, worse in another, any gains fragile and all too often only temporary. As I write the UK government is again talking of burying nuclear waste in the Snowdonia National Park, a few miles north of where I live. I was involved in the huge and successful resistance to that back in the 1980s and there is a weary feeling of *deja vu*, here we go again.

Given that we've messed things up, we should try to sort them out again, right? The underlying principle of a century of modern conservation is that humans have a moral and practical responsibility to work to reverse the damage that we have caused in ecosystems around the world. Pollution clean-ups, establishment of protected areas, anti-poaching measures, restoration ecology, control of invasive species, translocations, re-introductions and according to some conservation philosophers even recreation of species from stored genetic material are all aspects of our justifiable attempts to turn the clock back on our earlier misdeeds.

But some people question this approach − or at least the extent of this approach − and their ideas need to be considered seriously. The "re-wilding" movement aims to restore wild nature in places where this has almost disappeared, like Western Europe, shifting the baseline back to the Pleistocene. Some wilderness proponents also argue that if we recreate the right set of conditions we should not have to interfere any more but simply let nature take its course. Rewilding enthusiasts go further and argue that we need to step away from a defensive approach, which has run its course and alienates many people, and look at visionary new landscapes and seascapes, building an ecological vibrancy and balance not seen for thousands of years.

The originators of modern protected areas assumed that by protecting areas of land and water, nature would continue unimpeded. Today many protected areas are managed far more intensively than the average visitor notices: through prescribed burning, artificial control of water levels, creation or recreation of habitats designed for particular species, artificial nesting and roosting sites, culling of species when they become too numerous, removal of alien invasive species and sometimes even chemical control of things like water acidity. Do we need all this? If we simply left nature to get on with it, things might look very different for a while – might indeed look "untidy" – but perhaps they would settle into a different kind of balance. These are the arguments put forward also by those recognising the emergence of novel ecosystems[1]: as climate changes species will need to develop new assemblages. Some species will inevitably fall away, others will evolve over time as conditions continue to alter.

I don't think there is one way; if we take biodiversity rights seriously, we need to develop a portfolio approach, all the way from visionary rewilding to the protection of rump populations in zoos, botanical gardens and seed banks. Interventions will range from the management of tiny reserves aimed at protecting a single plant or imperilled insect to mega-marine reserves aimed at maintaining an entire ecosystem. They will involve state-run protected areas, traditional management in Indigenous peoples' territories, private initiatives, management of company-owned lands, careful rangeland management, Locally Managed Marine Areas maintained by fishing communities and marine protected areas managed by governments and a whole portfolio of other effective area-based conservation mechanisms. We need every tool in the toolbox.

I think, from the perspective of biodiversity rights, that some degree of stewardship is going to be needed, actually considerable stewardship, if we are not to suffer even more catastrophic losses of species than we are experiencing already. This doesn't reject the principle that the less we intervene the better, if only from the perspective of cost and effort. There should also be consideration about types and duration of such interventions: placing bat boxes in an area of woodland to rebuild population numbers is fine but should ideally be mirrored by efforts to manage the wood in such a way that natural bat roosts, in veteran trees for instance, emerge again in the future. Recreating a habitat that attracts a species is likely a safer option than translocation of a species from one habitat to another, although there will be circumstances when this is justifiable as well and these circumstances may increase in the future. Our actions today should always be within a strategy that looks towards self-sustaining ecosystems in the future.

SOME INTERVENTIONS

Various responses have been given, or at least hinted at, when discussing the tensions and challenges arising in deciding on and implementing biodiversity rights. In the next section, key points are summarised into what will hopefully be a single, digestible set of ideas.

As noted, biodiversity rights does not replace practical, subsistence or economic values of biodiversity but exists alongside these, both intrinsic and utilitarian rights need to be recognised and factored into conservation practice.

As humans, we have been the largest disruptive force on the planetary ecosystem for millennia, this era is rightly becoming known as the Anthropocene.[2] It is therefore a moral duty that we take all steps to halt and reverse the damage that we have caused; walking away is not an option.

All species and ecosystems have rights. From a practical perspective, we will need to prioritise the rights of some species and probably some ecosystems over others but these decisions need to be taken on a case-by-case basis. Cultural associations will play a role here, with some species favoured by reason of their cultural or sacred associations. But care is needed that decision are not driven by an unnecessarily narrow outlook, such as spending disproportionate efforts to conserve something at the edge of their range, when common elsewhere. It is vital not to lose sight of the importance of the obscure and unknown alongside charismatic keystone species.

"All species" includes all plant and fungi species. There are cases in addressing biodiversity rights when the rights of plants will exceed those of animals, for example if the survival of a plant species is being threatened by an introduced animal species.

Ecosystem rights are more complicated. Nevertheless, a combination of knowledge about the intricate interconnections within ecosystems that regulate and maintain the system, albeit with changes, and recognition of the common-sense reactions of many people in government and civil society, means that the claim that ecosystems deserve and should have rights respected can be made with some confidence.

It has become clear that animal rights sometimes conflict with biodiversity rights,[3,4] particularly but not only in cases of invasive species control and human-wildlife conflict (or perhaps better to say problems with human-wildlife coexistence). Trade-offs will be needed between these two philosophies and principles of action on occasion[5] and these may be hard to agree between all parties. Wherever possible animal rights should be taken into careful consideration,[6] for example in the control of problem animals, and efforts should be made to avoid unnecessary suffering or distress in conservation actions. Technological advances in management of human-wildlife conflict are making this easier all the time. Public reaction against the death of horses in cold weather in the Dutch rewilding experiment mentioned earlier suggest that the theory of letting nature take its course within human-managed conservation initiatives has strict limits in terms of acceptability to people.

Addressing potential tensions between human rights and biodiversity rights presents a major challenge for conservationists, human rights groups and others as we approach the second quarter of the twenty-first century. Many fine words and commitments have been made but they need to be implemented and potential pitfalls have so far been glossed over in most discussions. Human rights do not always trump biodiversity rights, for example if the human rights in question are the rights to development activities that put fundamental biodiversity rights under pressure. But we also need to recognise that preservation of biodiversity rights at the expense of the poorest and politically weakest members of society is unacceptable. Two key challenges are the ways in which protected areas and other effective area-based conservation measures are identified/recognised, agreed, planned and managed in ways that do not interfere with human rights, and ways of co-existence with animals that can pose threats to life or wellbeing.

The tools, agreements and processes to ensure that ambitious goals for area-based conservation do not undermine human rights are all generally in place, but often untested in many areas. Emergence of community-initiated protected areas and other places where biodiversity conservation is prioritised is a welcome sign that things can be done better. But there is a long way to go here and many practical problems still to fix.

From a biodiversity rights perspective, it seems clear that we should be eating less meat and dairy products overall and moving towards a much more plant-based diet; affordable substitutes including artificial meat are becoming more available all the time. If and when we do eat livestock products, these should be from sources where livestock is managed in a way that animals play a positive role in land management, such as substituting for extinct Pleistocene species. I'm aware that there are people who passionately disagree with this stance. And there are other considerations that also have indirect impacts on biodiversity such as the degree of processing and the transport involved in bringing food to your plate.

Trophy hunting is, paradoxically, not usually a problem from a biodiversity rights perspective, although it clearly has major implications for animal rights and often for human rights as well.

Rewilding is one significant approach to restoration, although careful reading of the rewilding literature does not convince me that it is more than one more step in an existing process; all the elements noted in rewilding have been suggested separately and often put to practice in more conventional approaches to ecosystem restoration. The word clearly has caught the zeitgeist with some people although it is also a divisive term with many farmers and with Indigenous peoples and other local communities who point out their long-term stewardship of places that outsiders refer to as "wilderness". I'm very nervous about the reintroduction of species "like" those that disappeared tens of thousands of years ago, given a long history of mismanaged introductions for biological control. But the general concept that we need to move from highly managed to more self-sustaining ecosystems in area-based conservation is correct. That said, in the medium term those highly managed reserves, populations in zoos and botanical gardens and seed bank repositories will all remain important tools for addressing the biodiversity crisis.

We probably spend too long talking about the wrong alien species. Many are clearly here to stay and there are more important things to worry about than their control; many invasive plants fall into this category. Ecosystems are fluid and adaptable and can generally absorb new components. Claims of the level of risk to species and the economic costs are often too high. From a biodiversity rights perspective, it clearly makes sense to invest in removing those alien species posing an immediate threat to existing endemic or range-limited species; this is classically the case in islands but may also be true in isolated water bodies for instance. Time and effort in controlling these, and thus allowing other species to survive and evolve, is important particularly as these isolated ecosystems often have unusual species evolved over long periods of time.

All of the above points suggest the need for greater synergy, between the actors discussed above related to the three great "rights" discussed in this book, but also with faith groups, commercial companies, governments, local communities, Indigenous peoples' groups and others. And we must all remain willing to learn, rethink, adjust; everyone is still very much on a learning curve here.

BRINGING BIODIVERSITY RIGHTS INTO FOCUS

However, this is not primarily a book about *how* to do conservation so much as about *why* we should be doing conservation. It started with a critique of what I believe is a narrow emphasis on the benefits of nature to people and a parallel disregard for the significance of nature for its own sake – nature's own rights. To be clear, benefits to people are real, almost always undervalued and underplayed by governments and they remain a critical part of the picture. But they are only one aspect of the conservation debate. Also, and something that many conservation bodies seem to have forgotten, they are not the reason why most people get involved in or show support for conservation in the first place. Over-emphasis on nature's benefits is risky; if we pin everything onto what we get out of nature, what happens if someone develops an artificial alternative that outperforms an ecosystem service?

This isn't a trivial question. I was involved in a couple of the reports from the TEEB project (The Economics of Ecosystems and Biodiversity), which was an early attempt to consolidate information about the economic importance of natural ecosystems.[7] Alongside some excellent examples of cases where calculation and publicity of economic benefits of nature led to better conservation there were others where pinning hopes on an economic analysis of ecosystem benefits was thoroughly undermined by a counter-analysis showing the greater benefits of sacrificing the ecosystem and allowing development. It is too late then to start shouting about intrinsic values. Once the issue has been narrowed down to a debate between economists then it is often the case that the group with the smarter economist will come out with the higher figure. And the forces of unsustainable development can often afford the smartest economists. In my experience, arguments about conservation are seldom won by cold facts alone.

One of the major changes – something fundamental enough to merit the over-used phrase a *paradigm shift* – has been the greater emphasis being put onto the rights of Indigenous peoples and local peoples in determining what happens on their traditional territories. Not to be naïve, these people will remain marginalised and under pressure in many places but it will for instance become more difficult for companies under public scrutiny to ignore or trample over their rights. And for conservation organisations to do the same. Once civil society get a fair chance at influencing decisions the question of intrinsic values become much more important and a potential tipping point in many cases.

We need to stop running scared of making a flat-out ethical case for conservation. We should not be put off by fears that people will think that this is tantamount to saying that people don't matter; the same people who are lobbying for human rights are almost unanimously aware of and concerned for something that approximates to biodiversity rights as well. Nor should we assume that these arguments will appear to be too wishy-washy or insubstantial to appeal to governments or industry either; ethical rights are in addition to rather than instead of issues like ecosystem services. While there are certainly plenty of politicians and industry leaders who demonstrate little concern for biodiversity there are others for whom an ethical argument holds important sway. Ethical issues and rights issues match well with all the main world religions and virtually all smaller faith groups; I would hazard a guess that the tendency to avoid such issues has increased the separation between faith and environment rather than the reverse.

We need an urgent debate on the whole issue of biodiversity rights, with three main aims:

1   To gain or regain a strong, public commitment to these rights from a range of organisations, including statements within the Convention on Biological Diversity, acknowledgement from conservation organisations and perhaps even more importantly from human rights bodies, Indigenous peoples' representatives, industry groups and others including those not usually associated with conservation. Signing on to a simple statement might be the quickest way to make progress.

2   To work through intersection between biodiversity rights and human rights, to identify potential clashes, with examples, and to suggest ways in which these might be resolved or at least accommodated. This might be addressed through meetings, online webinars, open forums or joint publications, with the key aims of bringing issues out into the open and searching for collaborative solutions. Some ombudsman role for assessing conservation projects has long been suggested and needs careful consideration.

3   To work through similar issues relating to biodiversity rights and animal rights. There is already a rich and well formulated set of arguments about animal rights or sentient rights; these need to be tested against some of the potential conflict points with biodiversity rights, such as control of invasive species (and trade-offs between different losses), whether plant rights ever trump animal rights and the tackling of human-wildlife conflict. Again, active debate with some clear objectives would be a good place to start.

The UN declaration on harmony with nature is an extremely important starting point, as are the various statements about Mother Earth. But the first doesn't focus particularly on rights and the second may just be too esoteric for a range of stakeholders; some faith groups will find the concept of Mother Earth hard to handle for example. I note that Mother Earth day on 22 April is not actually promoted or celebrated by many conservation organisations. That is a pity and I'd argue that it should be given greater prominence but it also suggests that something else is needed as well.

"Biodiversity" is a two-edged sword. I've already noted that it has a low level of recognition amongst the public and it doesn't necessarily play out well with some important stakeholders, like Indigenous peoples who see nature as such more holistic than simply biological diversity. When we were revising the IUCN definition of a protected area we shifted the emphasis from biological diversity to nature for exactly this reason.[8] But in this case our audience are not primarily Indigenous peoples, who already tend to have world views that embrace the rights of nature or biodiversity anyway. The main audiences for having a discussion about, and hopefully reaching agreement on, the critical importance of recognising the intrinsic rights of biodiversity are governments, non-governmental organisations both conservation and human rights, and companies. The people likely to be addressing these issues in these institutions are also likely to have a good understanding of what "biodiversity" means.

Alongside all the fine words about the environmental crisis and the benefits that biodiversity brings to humanity, we need agreement about the intrinsic value and the rights of biodiversity itself. Biodiversity was important before humans evolved and it will be important once we have disappeared, or ourselves evolved into something else. One way to do this would be to have a simple and clear statement that anyone – organisations, governments, multinational companies, church groups, wandering nuns, scout troops and sporting bodies – could sign onto. The following is my first attempt at what such a statement might look like.

A MANIFESTO FOR BIODIVERSITY RIGHTS

CONCERNED that global biodiversity is declining at a rate far above levels expected through natural evolutionary processes.

ACKNOWLEDGING that this extinction pulse is the result of human activity – particularly due to land use change, over-harvesting, climate change, and agricultural and industrial pollution – so that human influences are now the dominant agent of change throughout the world, leading many to refer to the present age as the Anthropocene.

AWARE of evidence that loss of biodiversity makes ecosystems less resilient to existing pressures such as those from climate change.

RECOGNISING the extent to which human society depends on natural and semi-natural ecosystems for a wide range of services, including those relating to food and water security, disaster risk reduction, health benefits and mitigation of climate change, leading to a high level of co-dependence.

APPRECIATING that many of these ecosystem services also have a considerable economic value.

MINDFUL also of the multiple intangible benefits from natural ecosystems, including a wide variety of cultural, aesthetic, spiritual and sacred values.

STRESSING the need for an increased commitment of effort, resources and expertise to halt and reverse the current loss of biodiversity, highlighted most recently in the agreement of the Global Biodiversity Framework by signatory states of the Convention on Biological Diversity, and drawing also on multiple other commitments including the UN Sustainable Development Goals and the UN Decade on Ecosystem Restoration.

REALISING that defence of biodiversity and the environment needs to be carried out in a way that is respectful also of human rights and equity, particularly relating to the rights of the Indigenous peoples and local communities who live in some of the most biodiverse places on the planet.

CONCERNED that biodiversity conservation is increasingly being presented only from the perspective of the value and services that biodiversity provides to humans, important though these are.

CONVINCED that this utilitarian view is out of step with the opinion and philosophy of many people, and with the mass of opinion within civil society.

SUPPORTIVE of previous work on defining rights for other species including the Universal Declaration of the Rights of Mother Earth, Earth jurisprudence and multiple efforts to promote the rights of nature.

We agree that:

All biodiversity – ecosystems, species and genetic variation within species – has intrinsic value separate from and existing alongside other more utilitarian values.

All biodiversity – ecosystems, species and genetic variation within species – has the right to continue their natural span of existence as or within a functioning ecosystem and we explicitly acknowledge these biodiversity rights.

Action that leads to extinction of species or destruction of all or most of particular ecosystems should be recognised as biocide and be illegal under international law.

Biodiversity rights are not the same as human rights or animal rights and there is the potential for all three of these to come have conflicting aims, in which case respectful negotiation is needed to reach an acceptable compromise position.

All biodiversity has value although from a pragmatic viewpoint species that play a critical role in maintaining ecosystems may need to have priority in conservation efforts in some cases.

Interpreting biodiversity rights is complex and an evolving process. Ecosystems are in flux, due to climate change, the artificial spreading of species around the world and other environmental factors, decisions about whether to aim to maintain "original" ecosystems or accept changing ecosystems will need to be made on a case-by-case basis and will inevitably be influenced by human cultural and social considerations.

The concept of biodiversity rights needs to be reflected more centrally in statements from all sectors of society, including international bodies, national governments, conservation and human rights NGOs, companies and civil society and Indigenous organisations.

This is far from an impossible dream. I'm encouraged, amongst the generally bad news globally, by how many people – often not centrally involved in the conservation movement – are already acknowledging the rights of nature. We need now to grow this movement.

NOTES

1 Hobbs, R.J., Arico, S., Aronson, J., Baron, J.S., Bridgewater, P., et al. 2006. Novel ecosystems: Theoretical and management aspects of the ecological world order. *Global Ecology and Biogeography* **15**: 1–7.

2 Bonneuil, C. and Fressoz, J.B. 2017. *The Shock of the Anthropocene*. Verso, London and New York.

3 Sagoff, M. 1984. Animal liberation and environmental ethics: Bad marriage, quick divorce. *Osgoode Hall Law Journal* **22**: 297–307.

4 Paquet, P.C. and Darimont, C.T. 2010. Wildlife conservation and animal welfare: Two sides of the same coin? *Animal Welfare* **19**: 177–190.

5 Perry, D. and Perry, G. 2008. Improving Interactions between animal rights groups and conservation biologists. *Conservation Biology* **22** (1): 27–35.

6 Ramp, D. and Bekoff, M. 2015. Compassion as a practical and evolved ethic for conservation. *Bioscience* **65** (3): 323–327.

7 Kettunen, M., Berghofer, A., Bouamrane, M., Bruner, A., Chape, S., et al. 2011. Recognizing the value of protected areas. In: ten Brink, P. (ed.) *The Economics of Ecosystems and Biodiversity in National and International Policy Making*. TEEB and Earthscan, London: 345–400.

8 Stolton, S. and Dudley, N. (eds.) 2009. *Defining Protected Areas: An International Conference in Almeria, Spain, May 2007*. IUCN, Gland.

12

AFTERWORD

We are working with a mixed group of local people on the edge of a national park in Colombia in 2017, talking about what benefits they saw from the area.[1] Much to our surprise, the local Indigenous leader is happy to show us where the main sacred sites were and he pointed them out on a map. Surprised because for many people sacredness is a secret for a particular group and it would be sacrilegious for them to point it out to a stranger, however warm their feelings towards that person might be. But everyone is different. We duly put stickers on the map and the park rangers now know which areas to avoid. But what he said next was more interesting. He said that of course the whole forest was sacred, and the particular sacred areas were a part of this rather than distinct entities. And when asked what he particularly valued about the protected area he did not mention the plants they collect or the fish they take from the river or even the sacred natural sites, but the view across the valley and the fact that he liked to stand there and look.

I like to stand and look too. Most days when writing this book, I've walked up above my house and looked across the river, Afon Dyfi, to the Snowdonia National Park. My round trip is a couple of miles and a climb of four or five hundred feet, depending on my route. There's almost always something to see; birds passing through, a new fungus poking up through the grass, the clouds scudding over the mountains, a touch of snow on the ridge opposite. Skirting the ridge, I can look down the valley and catch a glimpse of the sea, with sand dunes reaching into the estuary mouth and evening light shining on the water. I'm with the guy in Colombia, I enjoy the view.

DOI: 10.4324/9780429346675-12

This book has been about rights. But the whole concept of biodiversity rights is a human construct, so this is about us and our feelings as well. I hope I haven't taken the reader down too many esoteric rabbit holes, but the main aim has been to counter the strictly utilitarian approach to conservation that contains some very clear and important dangers. It is also naïve; time after time I have seen decisions made in favour of conservation not because of some fancy calculation of carbon value or economic value (though these affect decisions as well) but because people cared on an absolutely instinctive level. A colleague in Mexico told me about laboriously trying to explain the ecosystem service values of a proposed protected area to a government minister and getting nowhere, being met with a completely blank face, until the minister realised where he was talking about and said of course the area was absolutely beautiful, he played on the beach as a boy, and it definitely needed to be kept, now he got it..... We tend to over-think things sometimes. This mode of caring can be interpreted as spiritual, or cultural, or ethical or simply a gut feeling. Lots of people care.

Rights, feelings and values: three critical elements in conservation success. Over the past two decades the debate has become dominated by values, and often narrowly on economic values. I hope that this short book has provided a brief guide to the other side of the coin.

NOTE

1 We were using the Protected Area Benefits Assessment Tool, see: Ivanić, K.-Z., Stolton, S., Figueroa Arango, C.F. and Dudley, N. 2020. *Protected Areas Benefits Assessment Tool + (PA-BAT+): A Tool to Assess Local Stakeholder Perceptions of the Flow of Benefits from Protected Areas.* IUCN, Gland.

INDEX

Note: **Bold** page numbers refer to tables.